Angular Statistics

Angular Statistics

A. V. Dattatreya Rao

S. V. S. Girija

CRC Press
Taylor & Francis Group
Boca Raton London New York

CRC Press is an imprint of the
Taylor & Francis Group, an **informa** business
A CHAPMAN & HALL BOOK

CRC Press
Taylor & Francis Group
6000 Broken Sound Parkway NW, Suite 300
Boca Raton, FL 33487-2742

CRC Press is an imprint of Taylor & Francis Group, an Informa business

Printed on acid-free paper

International Standard Book Number-13: 978-0-367-03000-1 (Hardback)

Library of Congress Control Number: 2019952550

Visit the Taylor & Francis Web site at
http://www.taylorandfrancis.com

and the CRC Press Web site at
http://www.crcpress.com

Dedicated to

The Almighty—A. V. Dattatreya Rao

My beloved mother and father—S. V. S. Girija

Contents

Foreword

Classical statistical analysis deals with data that was expressed as numbers with reference to an origin, in one or more dimensions. Data on length, area, volume, age, and time are some examples of this nature. In some typical studies, the observations carry not only magnitude but a direction as in the case of the angle at which an object is located in an astronomical zone. It is called angular data or directional data and circular data is a special case of directional data when the data is measured in two dimensions (study of celestial objects), and spherical data is a generalization of angular data to higher dimension.

Angular Statistics is an *apt contemporary contribution* to the desk of data scientists who are confronted with handling directional data in problems of pattern recognition of real-time situations. Most of the universities offering a post-graduate program in statistics do not have formal training in this subdiscipline. It is time to introduce this directional data analysis as a course such as SQC, Reliability, Operations Research, or Econometrics.

This book can be considered as "content" on mathematical statistics of angular data. The flow of text starts with the nature and context of angular data, understanding and summarization of directional data and proceeds toward development of new models from classical statistical distributions to handle angular data.

There are nine chapters in this book focusing on the mathematics of angular data including statistical modeling and inference. The idea of angular descriptive statistics is introduced in a lucid way by the authors in the first chapter. The limitations of linear models to handle circular data are highlighted by the authors, and angular statistics is stated as a tool to fill the gap.

Various methods like wrapping, exponentiation, stereographic projection, and distributions with offset are discussed in Chapters 2 through 5.

New techniques like Rising Sun circular models, stereographic semicircular models, and axial models are explained in the context of Cauchy, Weibull, and beta distributions.

Methods of estimating the parameters of circular distributions and measuring the goodness of fit of models are discussed in Chapter 8. These tools help in choosing the best among "available models." A model used to describe the pattern at a given time may not be suitable to a future context, and the analyst has to select the right model by adapting to the dynamic changes in the background parameters, or else incorrect outcomes will be predicted. In this sense, the authors have highlighted some measures of goodness of fit like Watson's U^2 test, and the current software tools help in automatizing the procedures with the help of p-values.

Angular statistics is a subdiscipline like a super-specialty in Medicine or statistical process control in production engineering. The statistical science behind the angular data can be better understood in a domain-specific discussion. The authors have chosen to explain the models in the context of Shewhart-type control charts for process average. I am sure that these tools are applicable in medical imaging (radiology) and for analyzing the changing patterns of organs in two and three dimensions. They are also used in the analysis of spatial data such as the geographical positioning system (GPS) and other applications that form a crucial component of *data analytics.*

I have all appreciation to the authors for bringing out these specialized areas of statistics to the readers of general statistics.

K. V. S. Sarma
Former Professor of Statistics
Statistics Consultant
Sri Venkateswara Institute of Medical Sciences (SVIMS), Tirupati

Preface

Directional data arise in many scientific fields including Biology, Geology, Geography, Meteorology, Physics, Political Science, and Image Analysis. Examples of such applications of directional data are given in the much-acclaimed book titled *Directional Statistics* by Mardia and Jupp (2000). A set of observations on directions is referred to as directional data. In particular, directional data of two dimensions is called circular data. Directional data arise in the form of circular/semicircular/axial, symmetric/asymmetric, and uni/bimodal data in practical situations of varied fields listed above. For the purpose of modeling such kinds of data sets, the data scientists found that existing models were inadequate. As there is paucity of angular models, and to fill the gap, this book is designed to construct new angular models with the existing techniques and to develop new tools of constructing angular models with an application to control charts in angular models. This book will be useful for data scientists/researchers/research students of statistics and allied fields.

The celebrated book *Topics in Circular Statistics* by Jammalamadaka and Sen Gupta was the source of inspiration for us to start working in this new emerging area in 2003. Around that period, a few wrapped models were available for modeling circular data, though different methods of constructing circular models were cited in the above book, and we found that not much was done based on other construction procedures. At the same time, it came to our attention that there is paucity of suitable models to work on the above types of data sets. That was the motivating point to take up the present work.

With this backdrop, and with the encouragement of the CRC Press group of publishers, we wrote this book, *Angular Statistics*, consisting of nine chapters. Chapters 2 through 4 are devoted to present methodologies of wrapping, inverse stereographic projection and offsetting, respectively. Our endeavor is to present various circular models derived using these procedures. In addition to the above, we could develop new techniques of construction of angular models using Rising Sun, positive definite sequences, and a differential approach, and the models derived using these new techniques are included in Chapter 5. Besides, discretization of continuous circular models and its application to derive discrete circular models are presented.

When the semicircular data is to be analyzed, the existing practice is to double the angles to carry out the analysis based on circular statistics and back-transforming the results for the interpretation of semicircularity. Further, Jammalamadaka and Sen Gupta have defined axial distributions restricting a circular model to an arc of an arbitrary length for the data analysis of axial data. It was observed that when inverse stereographic projection was applied on a linear model spanning from 0 to ∞, it turned out to be a semicircular model automatically/extemporaneously, and when offsetting is adapted on the bivariate beta model, the resultant model happened to be an axial model automatically/extemporaneously. Therefore, Chapter 6 is devoted to constructing such extemporaneous angular models.

Some of the circular and semicircular models are extended to l-axial (arc) models. Also, the procedure of the Marshall–Olkin transformation is extended to angular models to construct asymmetric l-axial models, which are included in Chapter 7.

Having discussed various construction procedures of angular models and deriving them, the next logical step would be to use them for modeling by applying one of the inferential techniques called fitting of distributions. If several models are found to be good fits, then discriminating among good fits based on their relative performance for the choice of the best fit in the class is the essence of the penultimate chapter.

In the last chapter, on the lines of Laha and Gupta (2011), the idea of control charts for circular and semicircular models are presented as one of the applications of these angular models. MATLAB is used for implementing various computations in this project.

Acknowledgments

We are thankful to CRC Press for bringing out the present book, *Angular Statistics*. Particularly we thank Aastha Sharma and Shikha Garg for coordinating throughout while preparing the manuscript. Also, we thank the editorial team.

As many as six PhD theses were awarded in this area by Acharya Nagarjuna University, Guntur, under the guidance of authors. The present work was prepared on the foundations of the contents of the above theses. In view of this, we are very much pleased to acknowledge the support rendered by Dr. V. J. Devaraaj, Dr. Y. Phani, Dr. A. J. V. Radhika, Dr. Ch. V. Sastry, Dr. P. S. Subrahmanyam, and Dr. K. Srinivas.

We specially express our gratitude to Prof. (Retd) K. V. S. Sarma, Department of Statistics, S. V. University, Tirupati, for providing us with a valuable foreword to this work.

We vow a lot to Prof. Ashis Sen Gupta for initiating us to the area of circular statistics, Prof. Jammalamadaka S. Rao for encouraging us to work in this area and our teacher, Prof. (Retd) I. Ramabhadra Sarama, Department of Mathematics, Acharya Nagarjuna University for clarifying several intricate mathematical notions throughout this work. We thank Prof. (Retd) S. Krishna Sarma for his encouragement and blessings. We pay homage to Late Prof. V.L. Narasimham, Department of Statistics, Acharya Nagarjuna University, Andhra Pradesh, India, who always stood as source of inspiration in nurturing our research career.

Authors

Dr A. V. Dattatreya Rao retired as Professor of Statistics and served at Acharya Nagarjuna University, Guntur, from 1980 to 2017. He taught various subjects ranging from statistical inference, multivariate analysis, pattern recognition, simulation theory, and computer programming to mention a few. His areas of interest are pattern recognition, inference from incomplete data, and directional data analysis. He has guided 10 PhD theses. He has published more than 100 research articles in national and international journals. He delivered several invited talks at various conferences and UGC refresher courses and training programs. He is the life member of the Indian Society for Probability and Statistics (ISPS), Indian International Statistical Association (IISA), and Society for Development of Statistics. He is the fellow of AP Academy of Sciences. He is the recipient of the Best Teacher Award from the Andhra Pradesh Government in 2011.

Dr S. V. S. Girija is currently serving as a Professor in the Department of Mathematics, Hindu College, Guntur. Her areas of research interest are directional data analysis and numerical analysis and computing. She has published 60 research papers in national and international journals. She has guided five PhD scholars and one MPhil student. She is the life member of Indian Society for Probability and Statistics (ISPS), Indian International Statistical Association (IISA), and AP Mathematical Society. She is the associate fellow of AP Academy of Sciences. She is the recipient of the Best Teacher Award from the Andhra Pradesh Government in 2014.

1

Introduction to Angular Data and Descriptive Statistics

1.1 Introduction to Circular Data

The observations in several diverse scientific fields are "directions." Such directions arise in statistical problems where the data is in the form of angular measurements giving orientations or angles in the plane (two dimensional) or in space (three or more dimensional). A set of such observations on directions is referred to as "DIRECTIONAL DATA"; in particular, directional data of two dimensions is called "CIRCULAR DATA."

If the directional data consists of only two-dimensional observations, then it is called circular data. Since it deals with two-dimensional angular data, the words "angular data" and "circular data" are used interchangeably. Since the direction has no magnitude, these can be conveniently represented as points on the circumference of a unit circle centered at the origin or as unit vectors connecting the origin to these points.

Directional statistics is an important branch of statistics for analyzing and modeling angular data. Angular data have many novel and distinctive characteristics and challenges in terms of its treatment and statistical analysis. These special features make angular data analysis distinct from the standard linear statistical analysis.

Circular data arise in many scientific fields, including Biology, Geology, Geography, Meteorology, and Physics. Many examples of circular data coming from widely differing scientific fields are detailed in the book *Directional Statistics* by Mardia and Jupp (2000).

The frequently quoted examples for the angular data are the measurements obtained from the clock and the compass. Typical examples of angular data from a compass are the measurements of geographical directions, wind directions, orientations of birds and animals, orientation of geological phenomena, etc. By taking a calendar year as a one-year clock, the data, such as number of tourists visiting a country in a year or the number of cases of a specified disease diagnosed at a hospital, can be considered as angular data.

1.2 Applications and Nature of Angular Data Sets

Directional data arise in various natural, physical, and social sciences in different kinds of phenomena, such as circular/semicircular/axial, symmetric/asymmetric, and unimodal/bimodal data. The following are some of the applications of angular statistics and the methods of circular statistics in analyzing the angular data.

1.2.1 Long-Axis Orientations of Feldspar Laths (Semicircular) Data

Measurement of long-axis orientations of Feldspar laths: the data of 133 directional measurements of feldspar laths in basalt reported by Smith (1988) and presented by Fisher (1993) on page 240 (in degrees) are as follows (Figure 1.1):

176 162 49 174 174 49 54 63 59 61 66 104 97 58 121 5 178 3 168 0 18 39 140 63
55 170 169 37 152 73 53 176 72 170 113 56 87 161 164 21 50 6 59 140 54 64 56
38 61 143 51 144 148 44 60 98 86 145 38 168 39 134 68 57 129 68 132 82 54 119
131 50 93 160 127 124 65 108 52 61 86 37 132 83 163 58 144 29 80 172 144 138
10 45 137 11 145 103 69 124 54 121 1 39 111 153 13 5 5 107 104 39 133 36 63 4
21 51 30 52 90 143 13 50 109 12 170 5 14 91 132 121

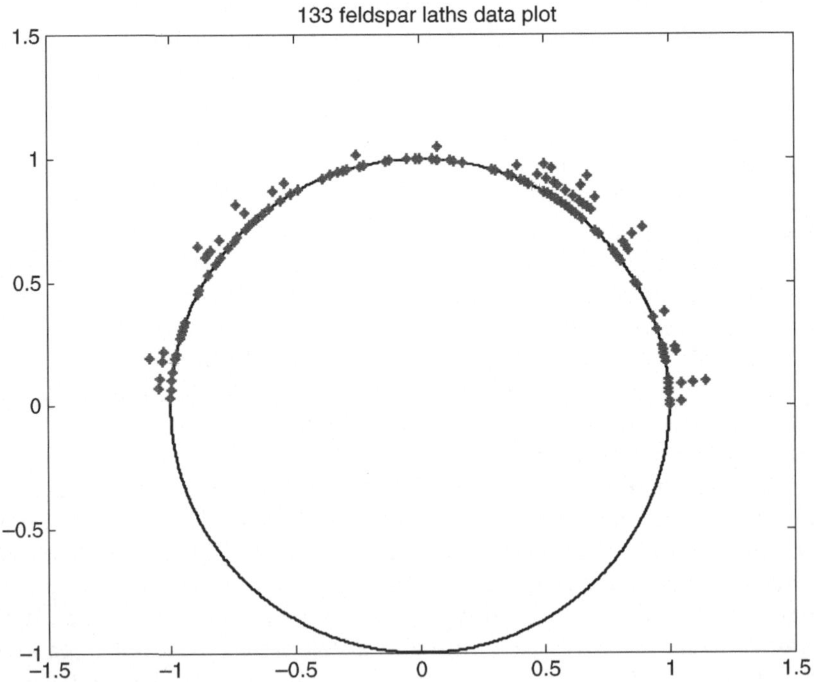

FIGURE 1.1
Graph of 133 Feldspar laths data plot.

1.2.2 Face-Cleat in a Coal Seam (Semicircular) Data

Sixty-three measurements of median directions of face-cleat (Fisher 1993, p. 254) from the Wallsend Borehole Colliery, NSW, Australia (in degrees) are as follows (Figure 1.2):

80 90 80 84 95 79 85 80 84 90 78 90 81 90 79 90 80 84 73 115 73 85 79 72 110
75 84 75 90 85 85 70 84 96 90 73 85 90 68 124 117 114 121 127 127 125 127 84
79 73 79 79 75 80 79 68 117 119 131 112 80 90 75

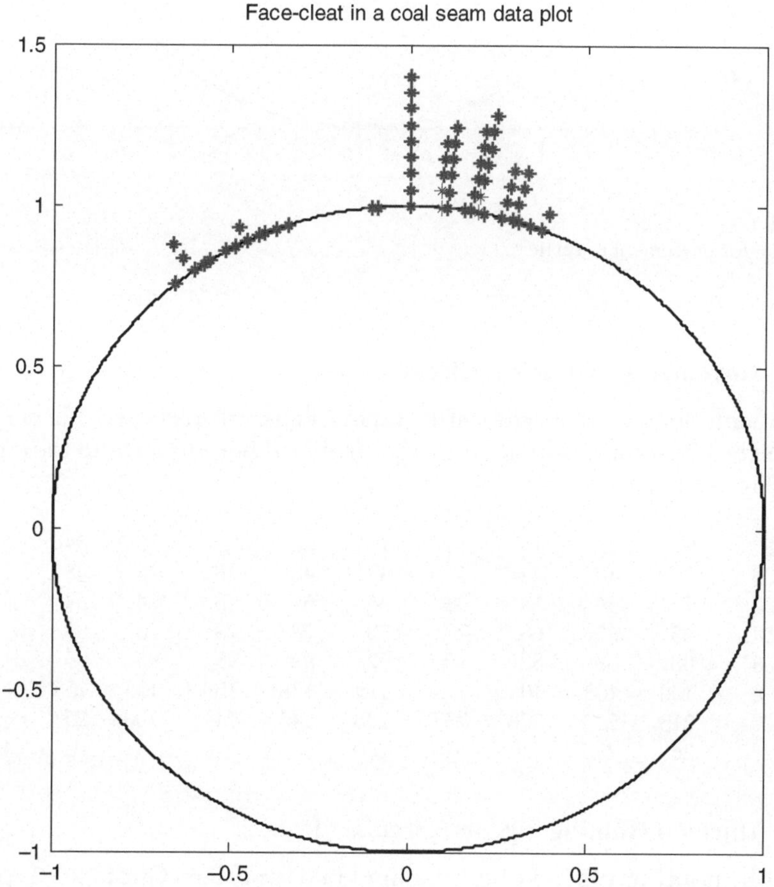

FIGURE 1.2
Graph of 63 face-cleat in a coal seam data plot.

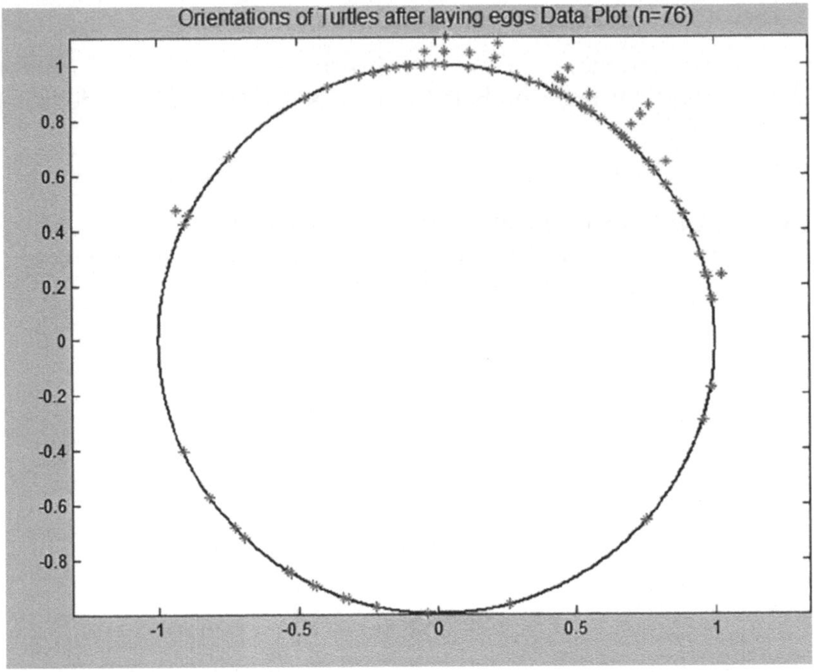

FIGURE 1.3
Data plot for movement of turtle data.

1.2.3 Movements of Turtles (Circular) Data

The orientations of 76 turtles after laying eggs are recorded (Fisher 1993, p. 241), the observations in degrees are furnished hereunder, and the circular data (Figure 1.3) are:

8	9	13	13	14	18	22	27	30	34
38	38	40	44	45	47	48	48	48	48
50	53	56	57	58	58	61	63	64	64
64	65	65	68	70	73	78	78	78	83
83	88	88	88	90	92	92	93	95	96
98	100	103	106	113	118	138	153	153	155
204	215	223	226	237	238	243	244	250	251
257	268	285	319	343	350				

1.2.4 Thirteen Homing Pigeon (Circular) Data

Thirteen homing pigeons (Jammalamadaka and Sen Gupta 2001, p. 165) were released singly in the Toggenburg Valley in Switzerland under subalpine conditions [data quoted in Batschelet (1981)]. They did not appear

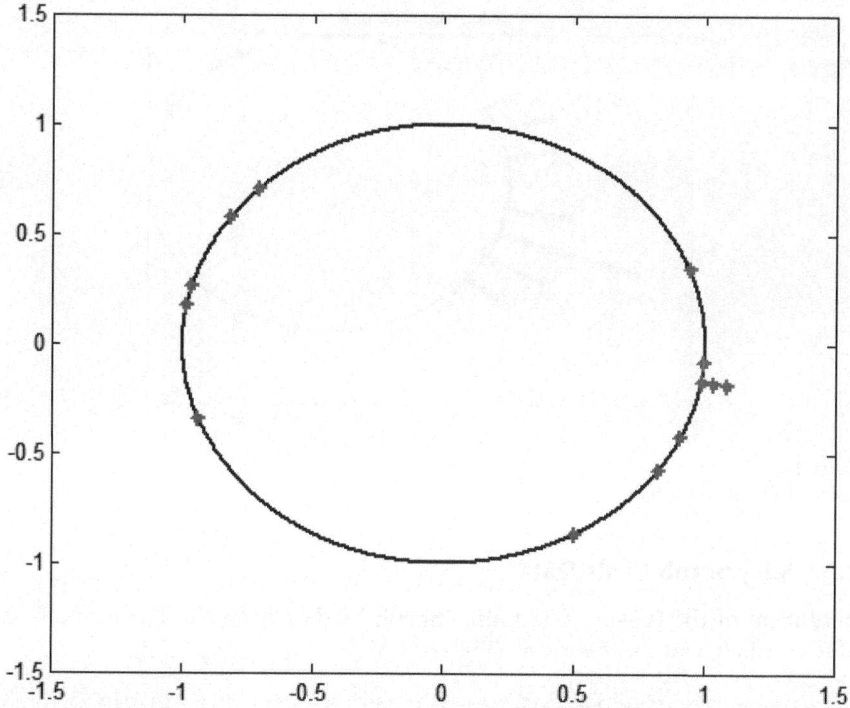

FIGURE 1.4
Graph of 13 homing pigeons data plot.

to have adjusted quickly to the homing direction but preferred to fly in the axis of the valley, indicating a somewhat **bimodal distribution**. The vanishing angles in degrees given in ascending order are as follows (Figure 1.4):

20, 135, 145, 165, 170, 200, 300, 325, 335, 350, 350, 350, 355

1.2.5 Movements of a Hundred Ants (Circular) Data

Directions chosen by a hundred ants (Fisher 1993, p. 243) in response to an evenly illuminated black target direction recorded in degrees (Figure 1.5) are as follows:

330 290 60 200 200 180 280 220 190 180 180 160 280 180 170 190 180 140 150
150 160 200 190 250 180 30 200 180 200 350 200 180 120 200 210 130 30 210
200 230 180 160 210 190 180 230 50 150 210 180 190 210 220 200 60 260 110
180 220 170 10 220 180 210 170 90 160 180 170 200 160 180 120 150 300 190
220 160 70 190 110 270 180 200 180 140 360 150 160 170 140 40 300 80 210
200 170 200 210 190

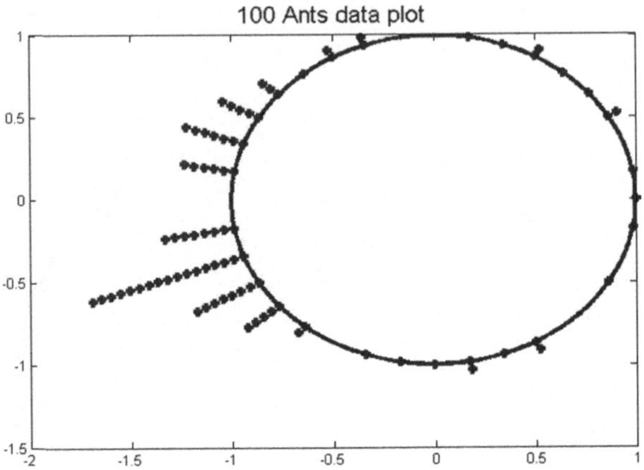

FIGURE 1.5
Graph of 100 ants data plot.

1.2.6 Noisy Scrub Birds Data

Orientation of the Nests of 50 Noisy Scrub Birds Along the Bank of a Creek Bed (Circular), Data in Degrees (Figure 1.6) (Fisher 1993, p. 252):

160, 145, 225, 230, 295, 295, 140, 140, 140, 205, 215, 135, 110, 240, 230, 250, 30, 215, 135, 110, 240, 105, 125, 125, 130, 160, 160, 105, 90, 130, 200, 240, 105, 125, 125, 125, 130, 160, 160, 250, 200, 200, 240, 240, 240, 250, 250, 250, 140, 140

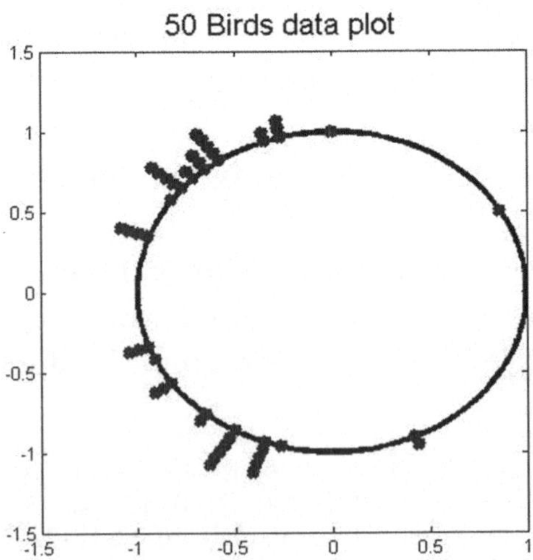

FIGURE 1.6
Graph of 50 birds data plot.

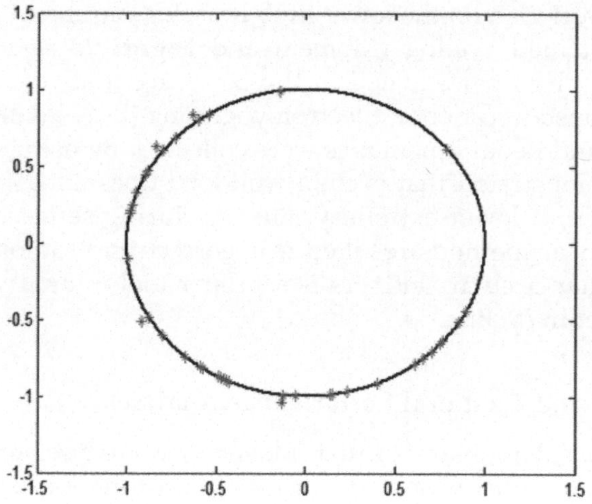

FIGURE 1.7
Graph of paleocurrents data plot.

1.2.7 Cross-Bed Azimuths of Palaeocurrents (Circular) Data

Set of cross-bed azimuths of Palaeocurrents measured in the Belford Anticline (New South Wales) (Figure 1.7) (Fisher 1993, p. 242):

284	311	334	320	294	137	123	166	143	127	244	243	152	
242	143	186	263	234	209	267	315	329	235	38	241	319	
308	127	217	245	169	161	263	209	228	168	98	278	154	279

Having presented various types of data sets, such as circular and semicircular, varieties of applications where angular data exist are presented.

1.2.8 Plant Phenology

Phenology is the study of periodic life cycle events of plants and animals and its relationship to climate (Patricia et al. 2010). Circular statistics applies well to phenological research to test the relationships between flowering time and other phenological or functional traits. Circular statistics also helps in grouping the species into annual, supra-annual, irregular, and continuous reproducers, to study seasonality in reproduction and growth and to assess the synchronization of species.

1.2.9 Study of Muscular Activity

Semireclined cycling is commonly recommended as a form of rehabilitation exercise. Techniques from circular statistics were used to investigate the effects of workload on the lower limb muscles' onset and offset patterns during semireclined cycling (Momeni and Faghri 2014). Fifteen healthy male, novice cyclists participated in three cycling trials of different workloads with a constant cadence. Electromyography (EMG) data of four lower limb muscles and pedal kinematics were collected. By using circular statistics, it was demonstrated that cycling workload does have an effect on the onset and offset of lower-extremity muscles during semireclined cycling, while using linear methods resulted in incorrect implications. These findings suggest that circular statistics is a proper tool in analyzing muscular onset and offset in cycling.

1.2.10 Spatial and Temporal Performance Analysis

Robert (1991) has demonstrated that linear-based analyses for circular data lead to erroneous conclusions and/or loss of information in the context of spatial and temporal performance measures. For example, the arithmetic mean and standard deviation are shown to be inappropriate descriptive measures for circular data. In addition, the use of average absolute deviations to measure directional errors of judgment can lead to loss of directional information. Finally, the usual methods of statistical inference are shown to fail in accounting for circularity when it exists. Therefore, the linear statistical methods are subject to serious, often unknown, and unrecognized errors in computing probabilities of Type 1 error, loss of statistical power, or both.

1.2.11 Study of Neurology

An application based on circular statistics was implemented to describe neuronal discharge patterns recorded during locomotion (Drew and Doucet 1991). Using this application, data was objectively classified with respect to both the mean direction and amplitude of their discharge, as well as to the variability (angular deviation) of that discharge. Also, the application described each cell's discharge as a single vector and provided an efficient method for comparing the discharge of a population of cells under two or more different conditions.

1.2.12 Political Science

Jeff and Dominik (2010) showed that circular data exists in political science and is generally mishandled by models that do not take into account the inherent recycling nature of some phenomenon. Hence, they applied the circular data analysis to political science and demonstrated that it is an appropriate statistical tool for political science data analysis. In this work, the authors highlighted the issues of circular data that are particular to political science and demonstrated how to construct models that treat these data appropriately. They developed a modeling framework based on the von Mises distribution and applied it to two data sets: casualties in the Second Iraq War and suicides in Switzerland. Results indicate significant improvement in data fit when properly accounting for the circular effects.

1.2.13 Psychophysical Research

Data arising from psychophysical studies requires careful processing due to its cyclical nature (Pearl et al. 2009). Unlike linear variables, such as response time and intensity, directions can be represented as angles or vectors on a circle, may have no natural zero-point or magnitude, and are defined on a periodic circle rather than an infinite line. Because of these unique features, directional data necessitates the use of circular statistical methods in psychophysical research. The implications of using linear statistics for circular data were explored by the authors in this work by submitting data from a behavioral study to both circular and linear statistical analysis.

1.2.14 Geography

Circular statistics can be used for the analysis of "journey to work" (JTW) data. The JTW data considered by Li et al. (2010) includes the total number of journeys between the origin (home) and destination zone (work) across South East Queensland (SEQ). Using bespoke tools developed in a GIS environment, the direction and frequency of each journey is first calculated. Using the outputs from this process, two descriptive measures—namely the circular mean and circular variance—are then computed and the results presented. Analysis of the mapped outputs shows strong JTW patterns that are readily detectable and visualized using a combination of GIS and circular statistics.

1.2.15 Archaeology

Several forms of archaeological data are best viewed in terms of circular statistics, including the orientations of settlements, dwellings, and graves (Bernard 2007). Numerous archaeologists have examined the orientations of graves with respect to various belief systems, the built environment, and aspects of the natural landscape. Their interpretations have been hampered by a reliance on conventional statistics.

1.2.16 Remote Sensing

Circular statistics is used for the analysis of errors in the positional accuracy of geometric corrections of satellite images using independent check lines (ICL) instead of independent check points (ICP). Circular statistics has been preferred because of the vectoral nature of the spatial error. The ground control point (GCP), ICP, and ICL data were acquired using differential GPS through field survey, and the plan metric positional accuracy was analyzed by both the conventional method (using ICP) and the proposed method (using ICL) (Cuartero et al. 2010).

1.2.17 Spatial Analysis

The positional error in spatial data is defined as a vector by comparing the coordinates between the true position and the measured position. The standard tests to assess the positional accuracy use only the magnitude of the vector and omit the azimuth; the use of both values allows a much more complete analysis of the positional error (Polo and Felicísimo 2010). Circular statistics tests are proposed that are relevant for this purpose and demonstrate that some important features are not identified by the common procedures. The test samples come from two data sets. The first is obtained from the comparison of 100 homologous points in two conventional maps, and the second one comes from the geometric calibration of a photogrammetric scanner. Error anisotropy is detected only by means of the circular statistics tests and density maps of distribution. Density maps working with both magnitude and angle can locate the outlier candidate and offer more information about the spatial distribution of error.

1.2.18 Plant Biology

Phenology is the study of recurring biological events and its relationship to climate. Nevertheless, the connection between the evaluation of temporal, recurring events and the analysis of directional data have converged and show circular statistics to be an outstanding tool by which to better understand plant phenology (Morellato et al. 2010). Circular statistics has particular value and application when flowering onset (or fruiting) occurs almost continuously in an annual cycle and importantly in southern climates, where flowering time may not have a logical starting point, such as mid-winter dormancy. Circular statistics applies well to phenological research where it is tested for relationships between flowering time and other phenological traits (e.g., shoot growth) or with functional traits such as plant height.

1.2.19 Kernel Regression for Directional Data

A method of kernel regression can be used to investigate the relationship between a directional explanatory variable and a real-valued response variable (Haiyong et al. 2011). Cross-validation and bootstrap methods for obtaining sensible bandwidths and standard-error estimates are also described. This method is very much used extensively in wildfire and meteorological data from Los Angeles County, California, with the goal of summarizing and quantifying the impact of wind direction on the total area burned per day in wildfires.

1.3 Limitations of Linear Statistics and Need for Angular Statistics

Directional data have many unique and novel features and challenges both in terms of modeling and in statistical treatment. For instance, the numerical representation of a two-dimensional direction as an angle or a unit vector is not necessarily unique since the angular value depends on the choice of the zero direction (origin) and sense of rotation. What is considered 60° by a mathematician who takes the east as the zero direction and anticlockwise as the positive direction comes out to be 30° to a geologist who takes the north as zero and clockwise as positive direction (called azimuth). It is therefore important to make sure that the conclusions of data analysis do not depend on the arbitrary choice of zero direction (origin) and the sense of rotation. The need for "INVARIANCE" of statistical methods and measures with respect to a choice of this arbitrary zero direction and sense of rotation makes many of the usual linear techniques and measures often misleading, if not entirely meaningless.

Commonly used summary measures on the real line, such as sample mean and variance, turn out to be inappropriate, as do all the moments and cumulants. Many notions such as correlation and regression as well as their statistical measures need to be revisited for directional data. Also, the notions of statistical inference as unbiasedness, loss functions, variance bounds, monotonicity of power functions of tests, etc. need to be redefined with caution.

1.4 Descriptive Statistics for Directional Data

As stated, the angular raw data can be represented in two ways. They can be represented by points on the circumference of a unit circle, the same being associated to each data point as shown in Figure 1.8. Alternately, they can be represented by drawing the radii of the unit circle, obtained by joining the origin to the data points on the circumference as in Figure 1.9.

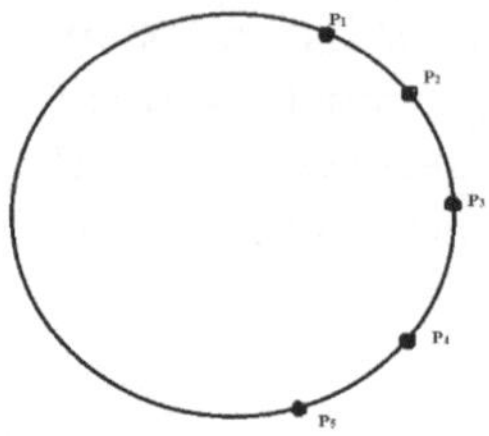

FIGURE 1.8
Points on a unit circle.

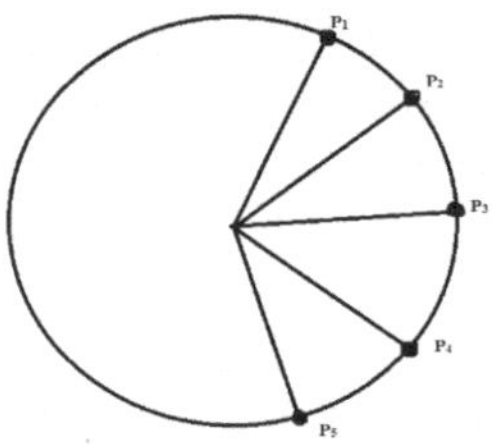

FIGURE 1.9
The radii of the unit circle.

Grouped data can be represented either by linear histograms, circular histograms, rose diagrams, or stem and leaf diagrams to gain an initial idea of important characteristics of the sample and suggest models for the data.

As in the linear case, the main emphasis will be on unimodal distributions. To describe unimodal circular distributions, some of the measures such as mean and variance are needed. These are useful in making comparisons between unimodal distributions. It is tempting to use the conventional measures on the line (linear data) for a circular distribution also. A drawback of such measures can be seen by considering an extreme example. Let us assume that the observed angles in a sample of size 2 are 4° and 356°. The arithmetic mean and the sample variance give absurd results. Although intuitively it can be inferred that the mean directions in some sense is 0°, and deviation about the mean is roughly 1°. The sensible answer is obtained by selecting the zero direction as the y-axis in place of x-axis, since the data then reduce to 266° and 274°. Hence, the usual linear statistical tools depend heavily on the choice of zero direction and are therefore inappropriate for circular distributions. Therefore, univariate descriptive statistics for circular data $\theta_1, \theta_2, \ldots, \theta_n$ are presented here.

1.4.1 Mean Direction

Let P_i be the point on the unit circle corresponding to the angle $\theta_i, i = 1(1)n$. Then the mean direction θ_0 of $\theta_1, \theta_2, \ldots, \theta_n$ is defined to be the direction of the resultant of the unit vectors $\overline{OP}_1, \overline{OP}_2, \ldots, \overline{OP}_n$. The Cartesian coordinates of P_i are $(\cos\theta_i, \sin\theta_i)$ so that the centre of gravity of these points is (C, S) where

$$C = \frac{1}{n}\sum \cos\theta_i \text{ and } S = \frac{1}{n}\sum \sin\theta_i$$

Then $R = \sqrt{C^2 + S^2}$ gives the length of the resultant vector. The direction θ_0 of the resultant vector

$$\theta_0 = \begin{cases} \tan^{-1}(S/C) & \text{if } S > 0,\ C > 0 \\ \tan^{-1}(S/C) + \pi & \text{if } \qquad C < 0 \\ \tan^{-1}(S/C) + 2\pi & \text{if } S < 0,\ C > 0 \end{cases}$$

is called the mean direction.

It may be noted that θ_0 is rotationally equivariant (if the data is shifted by a certain amount, the value of θ_0 also changes by the same amount), and similarly θ_0 has equivariant w.r.t. changes in the sense of rotation [when one switches from clockwise to anticlockwise, θ_i's become $(2\pi - \theta_i)$ and θ_0 will be $(2\pi - \theta_0)$].

1.4.2 Circular Distance and Measure of Dispersion

The direction of the resultant vector provides mean direction, and its length is a useful measure (for unimodal data) of "how concentrated" the data is toward this centre. If all the n observations (unit vectors) are pointing in the same direction indicating large concentrations, then its length can be as large as "n." Conversely, if the data is evenly spread over the circle indicating "no concentration," its length can be as small as zero. Therefore, the length of the resultant vector lies between $(0, n)$. Hence, $\bar{R}$ **the mean resultant length** associated with mean vector $(=R/n)$ lies in the range $(0, 1)$. Therefore, an appropriate measure of "distance" on the circle $(n–R)$ is indeed the right analogue of the usual sample variance. Following are other noteworthy measures of distance.

1.4.3 Distance between Any Two Points A, B on the Circle

Let A and B be any two points on the unit circle with respective angles θ_1 and θ_2 as shown in Figure 1.10. A reasonable measure of circular distance

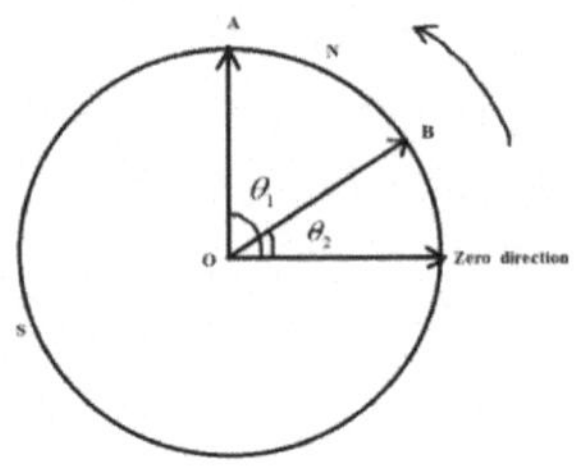

FIGURE 1.10
Measuring arc length.

between A and B is to consider the "smaller of the two arc lengths" between the points along the circumference.

Then the distance

$$d(A,B) = \min\left[(\theta_1 - \theta_2)\ ,\ 2\pi - (\theta_1 - \theta_2)\right] = \pi - \left|\pi - (\theta_1 - \theta_2)\right|.$$

Clearly, no two points on the circumference of a circle can be farther than π i.e., the circular distance lies in $[0,\ \pi]$. Alternatively, the distance can be taken as

$$d(\theta_1, \theta_2) = \left[1 - \cos(\theta_1 - \theta_2)\right]$$

which is a monotonically increasing function of $(\theta_1 - \theta_2)$ and lies in the interval [0, 2].

1.4.4 Circular Distance between Any Point to Several Points

If θ_i denotes the angle between u_i and an arbitrary point (vector) V $(0 \le \theta_i \le \pi)$ as shown in Figure 1.11, then the circular distance can be defined as $d(V, u_i) = [1 - \cos\theta_i]$.

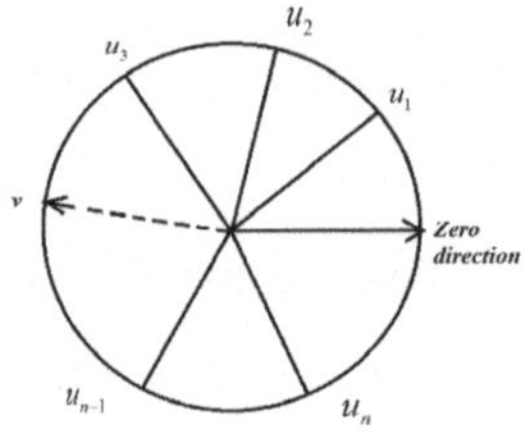

FIGURE 1.11
Points and arbitrary vector on circle.

Using this, the sample dispersion w.r.t. the arbitrary vector V, denoted by D_V, can be defined as

$$D_V(u_i, u_2, \ldots, u_n) = \sum_{i=1}^{n} d(V, u_i) = n - \sum_{i=1}^{n} \cos\theta_i$$

It can be observed that D_V attains its minimum when V is the normalized resultant vector $V^*\left(\frac{\sum\cos\theta_i}{R}, \frac{\sum\sin\theta_i}{R}\right)$.

1.4.5 Circular Distribution

In the continuous case $g:[0,\ 2\pi) \to \mathbb{R}$ is the probability density function of a circular distribution if and only if g has the following basic properties:

$$g(\theta) \geq 0, \quad \forall\theta \tag{1.4.1}$$

$$\int_0^{2\pi} g(\theta)\, d\theta = 1 \tag{1.4.2}$$

$$g(\theta) = g(\theta + 2k\pi) \tag{1.4.3}$$

for any integer k (i.e., g is periodic) (Mardia 1972).

It may be noted that the circular distribution (Jammalamadaka and Sen Gupta 2001) is a probability distribution whose total probability is concentrated on the unit circle $\{(\cos\theta, \sin\theta) / 0 \leq \theta < 2\pi\}$ in the plane, which satisfies the properties (1.4.1) through (1.4.3).

If $G(\theta)$ denotes the cdf of the r.v., the characteristic function of the circular model is given by

$$\varphi_\theta(t) = E\left(e^{it\theta}\right) = \int_0^{2\pi} e^{it\theta} dG(\theta) = \rho_t e^{i\mu_t} \quad t \in \mathbb{R}$$

It is known that whenever $\varphi(t) \neq 0,\ e^{2\pi it} = 1$ (Mardia 1972, p. 41). This suggests that the function $\varphi(t)$ should only be defined for integer values of t. Accordingly, the characteristic function $\varphi(p) = \varphi_p$ is defined by

$$\varphi_\theta(p) = E\left(e^{ip\theta}\right) = \int_0^{2\pi} e^{ip\theta} dF(\theta) = \rho_p e^{i\mu_p} \quad p \in \mathbb{Z}$$

Clearly, $\varphi_0 = 1,\ \bar{\varphi}_p = \varphi_{-p}$.

1.4.6 Trigonometric Moments

The value of the characteristic function φ_p at an integer p is called the pth trigonometric moment of θ. The real and the imaginary parts of φ_p are denoted by α_p and β_p, respectively. These trigonometric moments can also be viewed in terms of

$$\alpha_p = E(\cos p\theta),\ \beta_p = E(\sin p\theta),\ p \in \mathbb{Z} \tag{1.4.4}$$

The first trigonometric moment, namely, $\varphi_1 = \alpha_1 + i\beta_1 = \rho_1 e^{i\mu_1}$, plays a prominent role in determining the mean direction and resultant length (Jammalamadaka and Sen Gupta 2001).

Statistical analysis—modeling and inference form the basis for objective generalization from observed data. Some of the commonly used parametric models for directional data are constructed using one of the following methods:

1. Geometrical considerations
2. Wrapping a linear distribution around the unit circle
3. Characterization properties such as maximum entropy or maximum likelihood
4. Transforming a bivariate linear random variate to just its directional component, so-called offset distributions
5. Stereographic projection to a linear model
6. Applying Rising Sun function
7. By differential approach
8. Using the Marshall–Olkin transformation
9. Using Toeplitz Hermitian Positive Definite (THPD) matrices.

Discussion on the construction of angular models based on some of the above techniques forms the content of this book.

1.4.7 Population Characteristics of Circular Models

Mardia and Jupp (2000) gave expressions for **the mean direction, resultant length, circular variance, standard deviation, central trigonometric moments, skewness,** and **kurtosis** ($\mu_o, \rho_1, V_o, \sigma_o, \alpha_p^*, \beta_p^*, \gamma_1^o$, and γ_2^o) for circular distributions. These characteristics for the new circular models are also based on their respective trigonometric moments. These can be expressed in terms of trigonometric moments α_p and β_p.

1.4.8 The Mean Direction and the Resultant Length

The mean direction is given by

$$\mu_o = \mu_1^o = \tan^{-1}\left(\frac{\beta_1}{\alpha_1}\right) \tag{1.4.5}$$

In particular, the resultant length is

$$\rho = \rho_1 = \sqrt{\alpha_1^2 + \beta_1^2} \tag{1.4.6}$$

1.4.9 Circular Variance and Standard Deviation

The circular variance is

$$V_O = 1 - \rho \tag{1.4.7}$$

and the standard deviation is

$$\sigma_o = \sqrt{-2\ln(1 - V_o)} \tag{1.4.8}$$

1.4.10 Central Trigonometric Moments

The central trigonometric moments of a circular random variable X_W are defined by the equation

$$\mu_p = Ee^{ip(X_W - \mu_0)} = \alpha_p^* + i\beta_p^* \tag{1.4.9}$$

where μ_0 is the mean direction. Since the expectation in the above is given by $\varphi_p e^{-ip\mu_0}$, the central trigonometric moments are

$$\begin{aligned} \alpha_p^* &= \rho_p \cos(\mu_p^o - p\mu_o), \\ \beta_p^* &= \rho_p \sin(\mu_p^o - p\mu_o) \end{aligned} \tag{1.4.10}$$

1.4.11 Skewness and Kurtosis

The skewness and kurtosis of circular random variable X_S are expressed in terms of the second trigonometric moments. One common measure of skewness γ_1^o of a circular distribution is

$$\gamma_1^o = \beta_2^* / V_o^{3/2} = \frac{\beta_2\left(\rho^2 - 2\beta_1^2\right) - 2\alpha_1\alpha_2\beta_1}{\rho^2(1-\rho)^{3/2}} \tag{1.4.11}$$

The circular kurtosis is given by,

$$\gamma_2^o = \frac{\alpha_2^* - (1-V_o)^4}{V_o^2} \tag{1.4.12}$$

1.5 Existing Circular Models

The following are the circular models frequently used for modeling circular data till recent past:

a. **von Mises model**

The pdf of von Mises distribution (Jammalamadaka and Sengupta 2001) is

$$g(\theta) = \frac{1}{2\pi I_0(k)} \exp\left(k\cos(\theta - \mu)\right), \ \kappa > 0, \ \theta, \mu \in [0, 2\pi) \tag{1.5.1}$$

where I_0 denotes the modified Bessel function of the first kind and order zero, which can be defined by $I_0(k) = \frac{1}{2\pi}\int_0^{2\pi} e^{k\cos\theta} d\theta$.

The function I_0 has power series expansion

$$I_0(k) = \sum_{r=0}^{\infty} \frac{1}{(r!)^2}\left(\frac{k}{2}\right)^{2r}$$

The parameter μ is the mean direction, and the parameter k is known as the concentration parameter. The mean resultant length is $A(k)$ where A is the function defined by $A(k) = \frac{I_1(k)}{I_0(k)}$.

b. **Wrapped Cauchy model**

The probability density function of wrapped Cauchy distribution is defined as

$$g(\theta) = \frac{1}{2\pi} \frac{1-\rho^2}{1+\rho^2 - 2\rho\cos(\theta - \mu)} \tag{1.5.2}$$

where μ, σ are parameters such that $\rho = e^{-\sigma} (\sigma > 0)$ and $0 \le \mu < 2\pi$.

c. **Cardioid model**

The probability density function of Cardioid distribution is given by

$$g(\theta;\mu,\rho)=\frac{1}{2\pi}\left(1+2\rho\cos(\theta-\mu)\right) \tag{1.5.3}$$

where $\theta,\mu\in[0,2\pi)$, and $-\frac{1}{2}<\rho<\frac{1}{2}$

The distribution function of f is given by

$$G(\theta)=\frac{1}{2\pi}\left(\theta+2\rho\sin(\theta-\mu)+2\rho\sin\mu\right) \tag{1.5.4}$$

d. **Uniform distribution**

The probability density function of uniform distribution is given by

$$g(\theta)=\frac{1}{2\pi} \text{ where } \theta\in[0,2\pi) \tag{1.5.5}$$

e. **Wrapped normal (circular normal) distribution**

The probability density function of wrapped normal distribution is given by

$$g(\theta)=\frac{1}{\sigma\sqrt{2\pi}}\sum_{m=-\infty}^{\infty}\exp\left[\frac{-(\theta-\mu-2\pi m)^2}{2\sigma^2}\right] \tag{1.5.6}$$

where $\theta,\mu\in[0,2\pi)$ and $\sigma>0$

References

Batschelet, E. 1981. *Circular Statistics in Biology*, Academic Press, London.

Bernard, M. 2007. On the Application of Circular Statistics to Grave Orientations at Two Monongahela Village Sites, *Journal of Middle Atlantic Archaeology*, 23, 105–116.

Cuartero, A., Felicisimo, A.M., Polo, M.E., Caro, A. and Rodriguez, P.G. 2010. Positional Accuracy Analysis of Satellite Imagery by Circular Statistics, *Photogrammetric Engineering and Remote Sensing*, 76, 1275–1286.

Drew, T., and Doucet, S. 1991. Application of Circular Statistics to the Study of Neuronal Discharge During Locomotion, *Journal of Neuroscience Methods*, 38(2–3), 171–181.

Fisher, N.I. 1993. *Statistical Analysis of Circular Data*, Cambridge University Press, Cambridge, UK.

Haiyong, X., Kevin, N., Frederic, P.S. 2011. Kernel Regression of Directional Data with Application to Wind and Wildfire Data in Los Angeles County, California, *Forest Science*, 57(4), 343–352(10).

Jammalamadaka, S.R. and Sen Gupta, A. 2001. *Topics in Circular Statistics*, World Scientific Press, Singapore.

Jeff, G. and Dominik, H. 2010. Circular Data in Political Science and How to Handle It, *Summer*, 18(3), 316–336.

Li, T., Corcoran, J. and Burke, M. 2010. Investigating the Changes in Journey to Work Patterns for South East Queensland: A GIS Based Approach, Australasian Transport Research Forum 2010 Proceedings, http://www.patrec.org/atrf.aspx.

Mardia, K.V. 1972. *Statistics of Directional Data*. Academic Press, London, UK.

Mardia, K.V. and Jupp, P.E. 2000. *Directional Statistics*, John Wiley & Sons, Chichester, UK.

Momeni, K. and Faghri, P.D. 2014. Application of Circular Statistics in the Study of Muscular Activity During Semi-reclined Cycling, *40th Annual Northeast Bioengineering Conference (NEBEC)*, Boston, MA, pp. 1–2.

Morellato, L. P., Alberto, L.F., Hudson, I.L. 2010. Applications of Circular Statistics in Plant Phenology: A Case Studies Approach, *Phenological Research: Methods for Environmental and Climate Change Analysis*, Springer, Dordrecht, the Netherlands, pp. 339–359.

Patricia, C.L., Morellato, A.L.F. and Irene, L.H.. 2010. *Applications of Circular Statistics in Plant Phenology: A Case Studies Approach*, Springer, Dordrecht, the Netherlands.

Pearl S. Guterman, Allison, R.S. and Mc Cague, H. 2009. The Application of Circular Statistics to Psychophysical Research, *Proceedings of the 25th Annual Meeting of the International Society for Psychophysics*, 25(2009), 185–190.

Polo, M.-E. and Felicísimo, Á.M. 2010. Full Positional Accuracy Analysis of Spatial Data by Means of Circular Statistics, *Transactions in GIS*, 14, 4, 421–434.

Robert, P.M. 1991. Circular Statistical Methods: Applications in Spatial and Temporal Performance Analysis, Special Report 16, United States Army Research Institute for the Behavioral and Social Sciences.

Smith, N.M. 1988. Reconstruction of the Tertiary drainage systems of the Inverellregion. B.Sc.(Hons.) thesis, Department of Geography, University of Sidney, Sidney, Australia.

2

Wrapped Circular Models

2.1 Introduction

One of the biggest challenges that any analyst in statistics encounters in analyzing the data is to identify the probability distribution that describes the data taken up for the study in a better way. The more the knowledge one has about the underlying probability distribution, the better the statistical analysis and inference would be.

A considerable number of probability distributions exist in the universe of statistics, which are extensively used in data analysis. For example, the role of the Weibull distribution in reliability and quality control applications is immense. Exponential and Rayleigh distributions, which are special cases of the Weibull distribution, have many applications in life-testing data analysis. The inverted Weibull distribution is also one of the most popular probability distributions considered for analyzing the life-testing data with monotone failure rate. There are many more such popular distributions to add to this list.

Though these distributions play a key role in analyzing the data, in some of the cases, these distributions due to their inherent limitations are incapable of explaining the nature and characteristics of the data correctly. To overcome these challenges in modeling, many new distributions are developed using one or more existing distributions as a base. Some of such distributions are discussed herewith.

The generalized Pareto distribution was introduced by Pickands (1975). He extended the Pareto distribution by raising its cumulative distribution function to a positive power. Mudholkar et al. (1995) introduced the exponentiated Weibull distribution as a generalization of the standard Weibull distribution and applied this new distribution as a suitable model to the bus motor failure time data.

The generalized exponential distribution was introduced by Gupta and Kundu (2001). This distribution can be used as an alternative to the gamma or Weibull distribution in many situations. Nadarajah and Kotz (2006) developed the exponentiated Frechet, exponentiated Weibull, exponentiated gamma and exponentiated Gumbel distributions, as an extension to the Frechet, Weibull, gamma, and Gumbel distributions, respectively, on the same lines that of the generalized exponential distribution extend the exponential distribution. Akinsete et al. (2008) developed the beta–Pareto distribution. Silva et al. (2010) introduced the generalized exponential-geometric (GEG) distribution by raising the cumulative distribution function of an exponential geometric distribution to a positive power.

The gamma–Pareto distribution was developed by Alzaatreh et al. (2012, 2013) who also developed the Weibull–Pareto distribution. Kareema and Boshi (2013) developed the exponential Pareto distribution. Merovci and Puka (2014) generalized the Pareto distribution using quadratic rank transmutation.

To fit the inverted Weibull distribution for modeling the non-monotone failure rates life-testing data, Flaih et al. (2012) proposed a standard exponentiated inverted Weibull distribution by adding another shape parameter to the standard inverted Weibull distribution. Tahir et al. (2014) proposed a new three-parameter Weibull–Pareto distribution, which can produce the most important hazard rate shapes. Suleman and Albert (2015) presented another form of the Weibull–Pareto distribution called the new Weibull–Pareto distribution.

The lognormal distribution is commonly used to model the lives of units whose failure modes are of fatigue-stress nature. This distribution has been found applicable in several fields, such as Geology, Econometrics, Biochemistry, and Medicine, in particular, modeling the weights of children and found that it gives a good representation of flood flows. This distribution is also used extensively in reliability applications to model failure times.

This chapter deals with construction of new continuous and discrete wrapped circular models. The list of models constructed on the lines of Dattatreya Rao et al. (2007) are wrapped exponentiated inverted Weibull distribution (WEIWD), wrapped new Weibull–Pareto distribution (WNWPD), wrapped lognormal distribution (WLND), wrapped logistic distribution (WLGD), wrapped Weibull distribution (WWBD), wrapped extreme-value distribution (WEVD), wrapped binormal distribution (WBNRD), wrapped half logistic distribution (WHLD), wrapped binomial distribution (WBD), wrapped Poisson distribution (WPD), and wrapped logarithmic distribution (WLGRD). Having constructed the new continuous and discrete wrapped circular models, graphs of pdf, cdf, and characteristic functions of some of the above distributions are presented.

2.2 Methodology of Wrapping for Continuous Linear Models

If X is a random variable defined on $\mathbb{R}$, then the corresponding circular random variable X_W is defined by the modulo 2π reduction Girija (2010)

$$X_W \equiv X(\operatorname{mod} 2\pi)$$

It is clearly a many-valued function given by

$$X_W(\theta) = \{X(\theta + 2k\pi) / k \in \mathbb{Z}\}$$

If $f(x)$ is the probability density function (pdf) of the linear random variable X, then for the circular random variable X_W the corresponding pdf $g(\theta)$ is defined as,

$$g(\theta) = \sum_{k=-\infty}^{\infty} f(\theta + 2k\pi) \tag{2.2.1}$$

where $\theta \in [0, 2\pi)$.

It can be verified that $g(\theta)$ with total probability concentrated on the unit circle $\{(\cos\theta, \sin\theta) / 0 \le \theta < 2\pi\}$ in the plane satisfies all the properties from (1.4.1) to (1.4.3).

2.2.1 Proposition

The trigonometric moment of order p for a wrapped circular distribution Jammalamadaka and Sen Gupta 2001, p. 31 and Carslaw (1930) corresponds to the value of the characteristic function of the unwrapped random variable X, say $\phi_X(t)$ at the integer value p, i.e. $\varphi_p = \varphi_X(p)$.

Proof: Using the cdfs G and F of θ and X, respectively, we have

$$\varphi_p = \int_0^{2\pi} e^{ip\theta} dG(\theta)$$

$$= \sum_{k=-\infty}^{\infty} \int_{2\pi k}^{2\pi(k+1)} e^{ip\theta} dF(\theta)$$

$$= \int_{-\infty}^{\infty} e^{ipx} dF(x) = \varphi_X(p)$$

If α_p and β_p are the trigonometric moments and $\sum_{p=1}^{\infty}(\alpha_p^2+\beta_p^2)$ is convergent in the L^2 sense, then the random variable θ has a density g, which is defined *almost everywhere* by

$$g(\theta)=\frac{1}{2\pi}\sum_{-\infty}^{\infty}\varphi_p e^{-ip\theta}$$
$$=\frac{1}{2\pi}\left[1+2\sum_{p=1}^{\infty}(\alpha_p\cos p\theta+\beta_p\sin p\theta)\right],\theta\in[0,2\pi),p\in\mathbb{Z} \quad (2.2.2)$$

is the Fourier representation of circular model.

As already stated, the convergence of the sum is in the L^2 sense:

$$\lim_n\int_0^{2\pi}\left|g(\theta)-1-2\sum_{p=1}^{n}\left(\alpha_p\cos p\theta+\beta_p\sin p\theta\right)\right|^2 d\theta=0$$

From the uniqueness theorem, the distribution function G on the circle is uniquely determined by characteristic function (c.f. Mardia 1972, p. 80). The characteristic function and cdf of the resultant wrapped circular model are

$$\varphi_W(p)=E(e^{ip\theta})=\varphi(p),\quad p\in\mathbb{Z}$$

and

$$G(\theta)=\int_0^{\theta}g(\theta')d\theta',\theta\in[0,2\pi)$$
$$=\sum_{k=-\infty}^{\infty}\left[F(\theta+2k\pi)-F(2k\pi)\right] \quad (2.2.3)$$

Now, applying this methodology, the proposed new distributions have been derived in the subsequent sections.

2.3 Wrapped Exponentiated Inverted Weibull Distribution (WEIWD)

The exponentiated inverted Weibull distribution (EIWD) is a generalization of the inverted Weibull distribution by adding a new shape parameter $\lambda \in R^+$ by exponentiation to the inverted Weibull distribution function.

A linear random variable X is said to follow a two-parameter EIWD, if the distribution function of X takes the following form

$$F(x) = \left(e^{-x^{-c}} \right)^{\lambda} \tag{2.3.1}$$

where c and λ both are shape parameters and $0 < x < \infty$ and $c > 0, \lambda > 0$.

Hence, the probability density function of EIWD is

$$f(x) = \lambda\ c\ x^{-(c+1)} \left(e^{-x^{-c}} \right)^{\lambda} \tag{2.3.2}$$

where $0 < x < \infty$ and $c > 0, \lambda > 0$.

If $\lambda = 1$, this EIWD becomes the standard inverted Weibull distribution, and if $c = 1$, EIWD represents the standard inverted exponential distribution.

2.3.1 Probability Density Function (pdf) of Wrapped Exponentiated Inverted Weibull Distribution (WEIWD)

Applying the method of wrapping explained in Section 2.2, the pdf, $g(\theta)$ of the WEIWD is obtained as

$$g(\theta) = \sum_{k=0}^{\infty} \lambda c(\theta + 2\pi k)^{-(c+1)} \left(e^{-(\theta+2\pi k)^{-c}} \right)^{\lambda} \tag{2.3.3}$$

where $\theta \in (0, 2\pi), c > 0,$ and $\lambda > 0$.

The graph depicting the linear and circular representations of the pdf of WEIWD for different values of parameter c keeping the parameter λ constant at 2.0 are as follows (Figures 2.1 and 2.2):

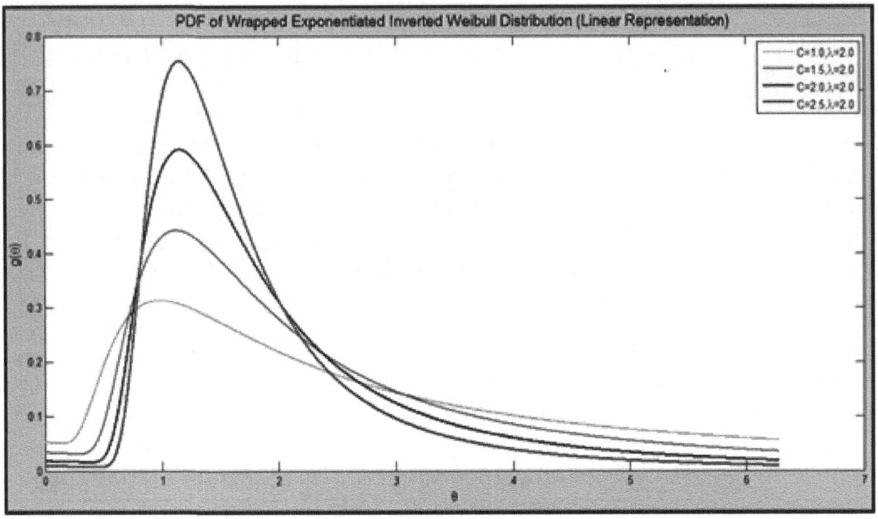

FIGURE 2.1
Pdf of WEIWD (linear representation for different values of *c*).

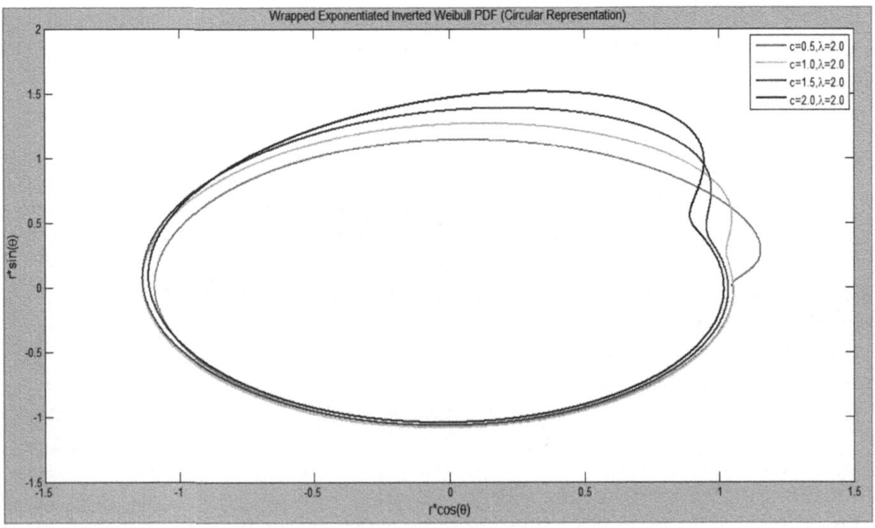

FIGURE 2.2
Pdf of WEIWD (circular representation for different values of *c*).

2.3.2 Cumulative Distribution Function (cdf) of WEIWD

The cdf, $G(\theta)$ of the WEIWD, is derived as

$$G(\theta) = \int_0^\theta g(\theta)d\theta.$$

$$= \int_0^\theta \sum_{k=0}^{\infty} \lambda c(\theta+2\pi k)^{-(c+1)} \left(e^{-(\theta+2\pi k)^{-c}}\right)^\lambda d\theta \tag{2.3.4}$$

$$G(\theta) = \sum_{k=0}^{\infty} (e^{-\lambda(\theta+2\pi k)^{-c}} - e^{-\lambda(2\pi k)^{-c}}) \tag{2.3.5}$$

where $\theta \in (0, 2\pi)$ and $c > 0, \lambda > 0$.

At different values for parameters c and λ, the graph of cdf for WEIWD is obtained as below (Figure 2.3).

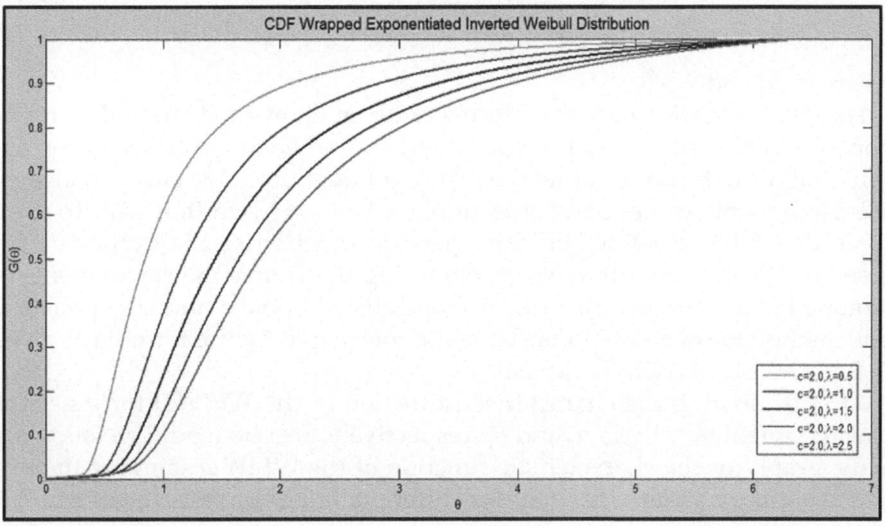

FIGURE 2.3
Cdf of WEIWD.

2.3.3 Characteristic Function of WEIWD

The characteristic function of the wrapped Exponentiated Inverted Weibull distribution is given by

$$\begin{aligned} \varphi_w(p) &= \int_0^{2\pi} e^{\,ip\theta}\, g(\theta)\, d\theta \\ &= \int_0^{2\pi} e^{\,ip\theta} \left(\sum_{k=0}^{\infty} \lambda\, c(\theta+2\pi k)^{-(c+1)} \left(e^{-(\theta+2\pi k)^{-c}}\right)^{\lambda} \right) d\theta \quad \text{for } p \in \mathbb{Z} \end{aligned} \tag{2.3.6}$$

To obtain trigonometric moments for WEIWD, Equation (2.3.6) has to be evaluated and is achieved from the characteristic function of EIWD using Proposition 2.2.1.

$$\begin{aligned} \phi_X(t) &= \int_0^{\infty} e^{\,itx} f(x)\,dx, \quad t \in \mathbb{R} \\ &= \int_0^{\infty} e^{itx} \lambda c x^{-(c+1)} (e^{-x^{-c}})^{\lambda}\, dx \end{aligned} \tag{2.3.7}$$

$$= \sum_{k=0}^{\infty} \frac{\left(it\lambda^{1/c}\right)^k}{k!} \Gamma_{(1-k/c)} \tag{2.3.8}$$

where $c > 0$ and $\lambda > 0$.

The characteristic function of the exponentiated inverted Weibull distribution is feasible only when the right-hand side of Equation (2.3.8) is convergent. But it can be noticed that the series in Equation (2.3.8) fails to converge at least for some values of c. For example, when $c = \frac{1}{n}$, $n > 0$, $n \in \mathbb{Z}^{+}$. To overcome this for evaluating the characteristic function of the exponentiated inverted Weibull distribution, for obtaining the trigonometric moments of wrapped exponentiated inverted Weibull distribution, a numerical integration method known as n-point Gauss–Laguerre quadrature formula as given in Rao and Mitra (1975) is applied.

After evaluating the characteristic function of the WEIWD for $p \in \mathbb{Z}$, the real and imaginary parts α_p and β_p, respectively, are obtained. The following is the graph for the characteristic function of the WEIWD showing the real and imaginary parts separately for different values of parameters c and λ (Figure 2.4).

2.3.4 Population Characteristics of WEIWD

The population characteristics for the WEIWD for some arbitrary values of the parameters c and λ are computed and tabulated (Subrahmanyam et al. 2017a). Similar tables can as well be generated for other values of the parameters.

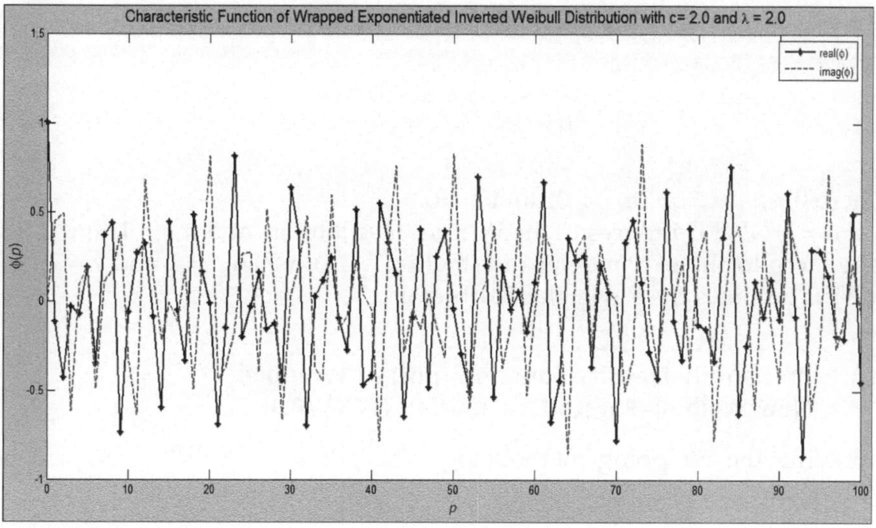

FIGURE 2.4
Characteristic function of WEIWD at $c = 2$ and $\lambda = 2$.

From the population characteristics, it can be observed that with increasing value of shape parameter c, keeping the other shape parameter $\lambda = 2$, the circular variance gradually decreases, the distribution is negatively skewed and remains platykurtic. With increasing value of the shape parameter λ keeping the shape parameter $c = 2$, the circular variance gradually increases, and the distribution starts shifting from negatively skewed to near symmetric and from platykurtic to mesokurtic.

2.4 Wrapped New Weibull–Pareto Distribution (WNWPD)

A linear random variable X is said to follow a three-parameter new Weibull–Pareto distribution (NWPD) if the distribution function of X is given by (Subrahmanyam 2017)

$$F(X) = 1 - e^{-\delta\left(\frac{x}{\lambda}\right)^{C}} \tag{2.4.1}$$

where $0 < x < \infty$, $c > 0$, $\lambda > 0$, and $\delta > 0$.

In NWPD, c is the shape parameter, and λ and δ are the scale parameters.

The pdf of the new Weibull–Pareto distribution is

$$f(x) = \frac{c\delta}{\lambda}\left(\frac{x}{\lambda}\right)^{c-1} e^{\left(-\delta\left(\frac{x}{\lambda}\right)^{c}\right)} \tag{2.4.2}$$

where $0 < x < \infty$, $c > 0$, $\lambda > 0$, and $\delta > 0$.

If $\delta = 1$, NWPD reduces to the Weibull distribution, and if $\delta = 1$ and $c = 1$, it reduces to the exponential distribution. Also, when $\delta = 1$ and $c = 2$, the NWPD becomes the Rayleigh distribution.

2.4.1 Probability Density Function (pdf) of Wrapped New Weibull–Pareto Distribution (WNWPD)

Applying the wrapping methodology, the pdf for WNWPD, $g(\theta)$ can be obtained as

$$g(\theta) = \sum_{k=0}^{\infty} \frac{c\delta}{\lambda}\left(\frac{(\theta+2k\pi)}{\lambda}\right)^{c-1} e^{-\delta\left(\frac{(\theta+2k\pi)}{\lambda}\right)^{c}} \tag{2.4.3}$$

where $\theta \in (0, 2\pi)$, $c > 0$, $\lambda > 0$, and $\delta > 0$.

The linear and circular representations of pdf for different values of δ keeping the values of c and λ constant at 2 are obtained as below (Figures 2.5 and 2.6).

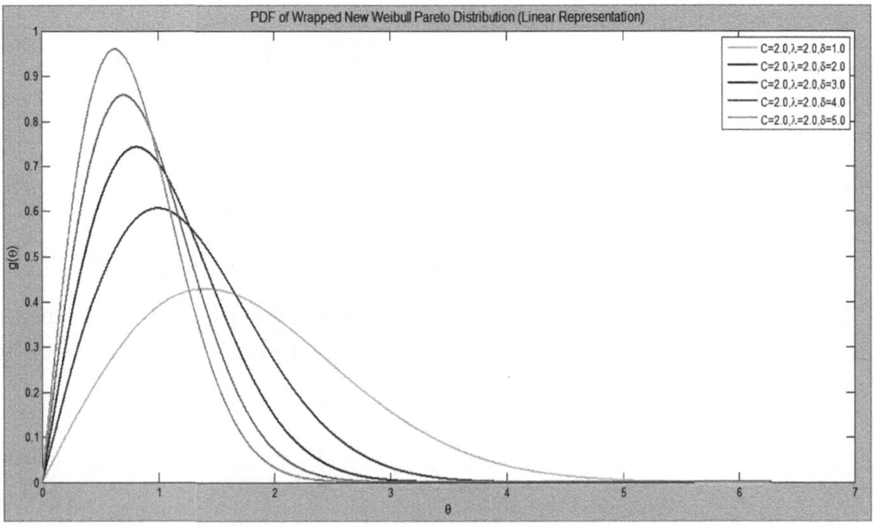

FIGURE 2.5
Pdf of WNWPD (linear representation for different values of δ).

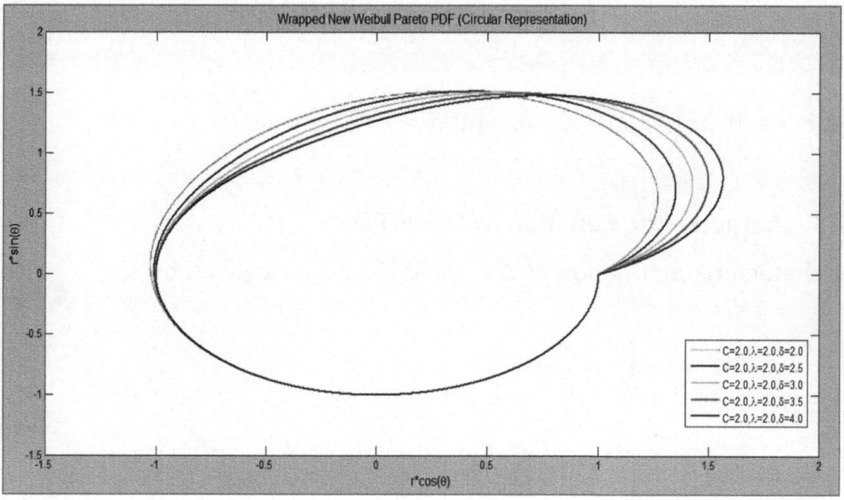

FIGURE 2.6
Pdf of WNWPD (circular representation for different values of δ).

2.4.2 Cumulative Distribution Function (cdf) of WNWPD

The cdf, $G(\theta)$ of the WNWPD, is (Figure 2.7)

$$G(\theta) = \int_0^{\theta} \sum_{k=0}^{\infty} \frac{c\,\delta}{\lambda} \left(\frac{(\theta + 2k\pi)}{\lambda} \right)^{c-1} e^{-\delta \left(\frac{(\theta+2k\pi)}{\lambda} \right)^{c}} d\theta \tag{2.4.4}$$

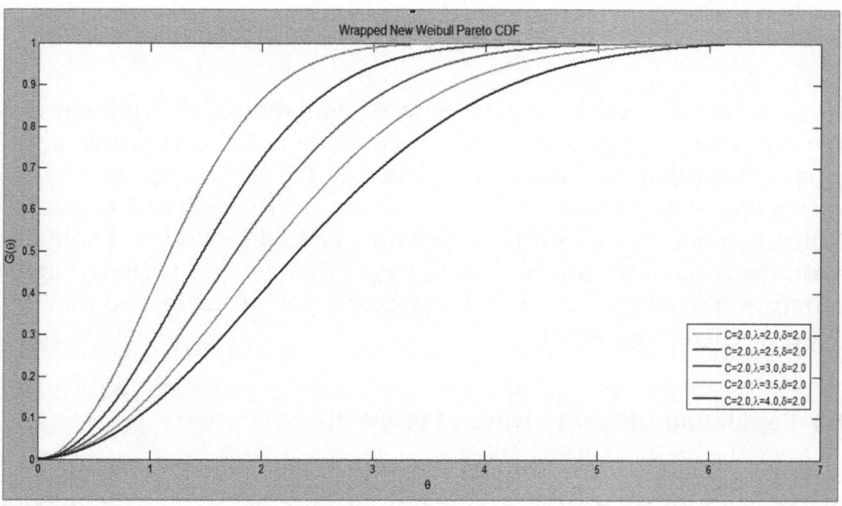

FIGURE 2.7
Graph of cdf of WNWPD.

$$= \sum_{k=0}^{\infty} \left(e^{-\delta(\frac{2k\pi}{\lambda})^c} - e^{-\delta(\frac{\theta+2k\pi}{\lambda})^c} \right) \tag{2.4.5}$$

where $\theta \in (0,2\pi),\ c>0,\ \lambda>0,$ and $\delta>0$.

2.4.3 Characteristic Function of WNWPD

The characteristic function of the WNWPD is hence given by

$$\varphi_w(p) = \int_0^{2\pi} e^{ip\theta} g(\theta)d\theta \quad \text{for } p \in \mathbb{Z}$$

$$= \int_0^{2\pi} e^{ipx} \left(\sum_{k=0}^{\infty} \frac{c\delta}{\lambda} \left(\frac{(\theta+2k\pi)}{\lambda} \right)^{c-1} e^{-\delta\left(\frac{(\theta+2k\pi)}{\lambda}\right)^c} \right) d\theta \tag{2.4.6}$$

Similar to the case of WEIWD discussed in Section 2.3, the characteristic function of NWP distribution can be evaluated for obtaining trigonometric moments for WNWPD.

$$\phi_X(t) = \int_0^{\infty} e^{itx} \frac{c\delta}{\lambda} \left(\frac{x}{\lambda} \right)^{c-1} e^{-\delta\left(\frac{x}{\lambda}\right)^c} dx \tag{2.4.7}$$

$$= \sum_{k=0}^{\infty} \left(\frac{(it\lambda)^k}{\delta^{(k/c)} k!} \right) \Gamma(1+k/c) \tag{2.4.8}$$

The characteristic function of the new Weibull–Pareto distribution is feasible only when the right-hand side of Equation (2.4.8) is convergent. But it can be noticed that the series in Equation (2.4.8) fails to converge at least for some values of c where $c = 1/n,\ n>0,\ n \in \mathbb{Z}^+$. To evaluate the characteristic function of the wrapped new Weibull–Pareto distribution for obtaining the trigonometric moments α_p and β_p for $p \in \mathbb{Z}$, the n-point Gauss–Laguerre quadrature formula for numerical integration as given in Rao and Mitra (1975) is applied (Figure 2.8).

2.4.4 Population Characteristics of WNWPD

The population characteristics of the WNWPD for computation of the characteristics for some arbitrary values the parameters are computed and tabulated (Subrahmanyam et al. 2017b). Similar tables can be generated for other values of the parameters.

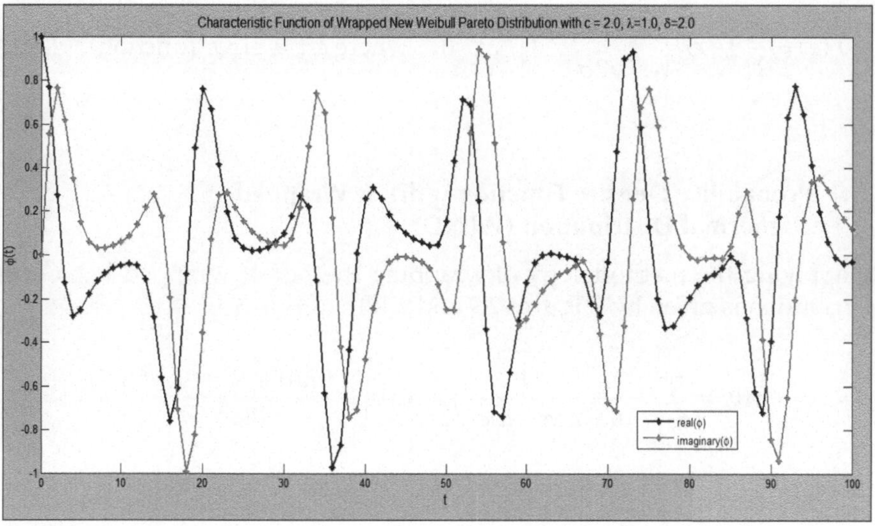

FIGURE 2.8
Graph of the characteristic function of WNWPD at $c = 2,\ \lambda = 1$ and $\lambda = 2$.

From the population characteristics, it can be noted that with increasing value of shape parameter c keeping both scale parameters $\delta = 2.0$ and $\lambda = 3.0$, the circular variance gradually decreased, the distribution incrementally reduced its negative skewness and remained platykurtic. With increasing value of scale parameter δ keeping other scale parameter λ at 2.0 and shape parameter c at 3.0, the circular variance gradually decreased, the distribution remained negatively skewed and platykurtic. Also with increasing value of scale parameter λ keeping other scale parameter $\delta = 3.0$ and the shape parameter $c = 3.0$, the circular variance gradually increased, the distribution reduced its negative skewness and remained platykurtic.

2.5 Wrapped Lognormal Distribution (WLND)

A continuous random variable X with density function (Johnson et al. 2000, p. 207)

$$f(x) = \begin{cases} \dfrac{1}{(x-\mu)\sigma\sqrt{2\pi}} \exp\left(-\dfrac{\{\ln(x-\mu)\}^2}{2\sigma^2}\right), & x > \mu, \quad \sigma > 0 \text{ is a parameter} \\ 0, & x \le \mu, \end{cases} \tag{2.5.1}$$

is said to have lognormal distribution with scale parameter $\sigma > 0$, and its distribution function is given by

$$F(x) = \frac{1}{2}\left[1 + erf\left(\frac{\ln(x-\mu)}{\sqrt{2}\sigma}\right)\right], \ \sigma > 0 \quad \text{where } erf \text{ is error function} \tag{2.5.2}$$

2.5.1 Probability Density Function (pdf) of Wrapped Lognormal Distribution (WLND)

By applying the methodology of wrapping the pdf of wrapped lognormal distribution is given by (Figures 2.9 and 2.10)

$$g(\theta) = \sum_{k=0}^{\infty} \frac{1}{(\theta + 2k\pi - \mu)\sigma\sqrt{2\pi}} \exp\left(-\frac{\{\ln(\theta + 2k\pi - \mu)\}^2}{2\sigma^2}\right),$$

where:

$$\theta, \mu \in [0, 2\pi), \theta > \mu, \sigma > 0 \tag{2.5.3}$$

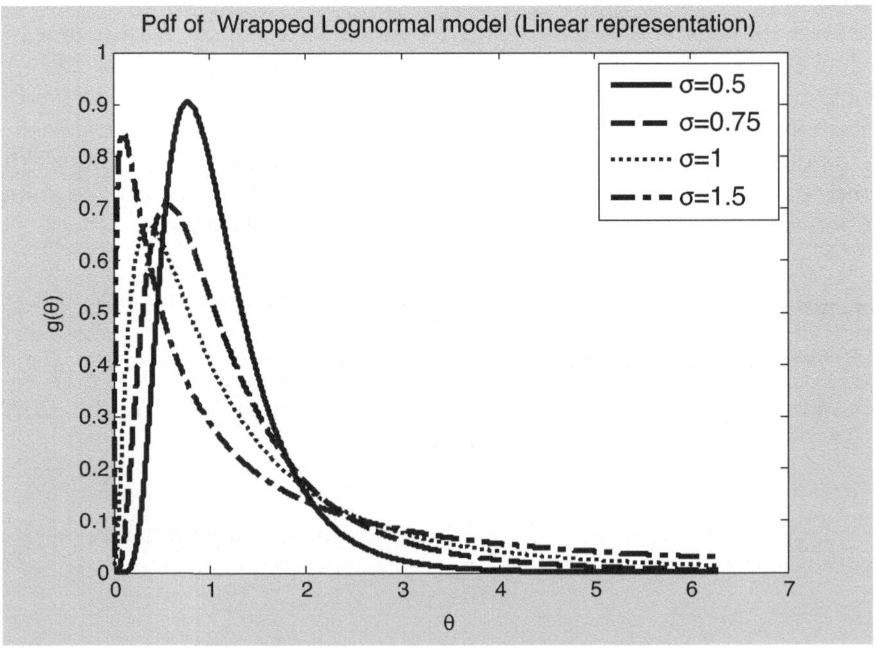

FIGURE 2.9
Graph of pdf of wrapped lognormal distribution (linear representation).

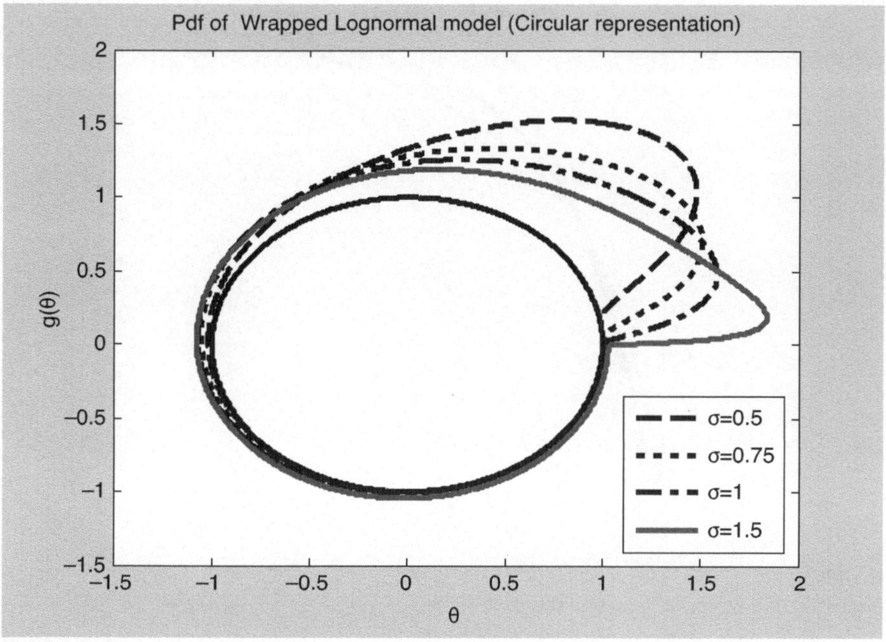

FIGURE 2.10
Graph of pdf of wrapped lognormal distribution (circular representation).

2.5.2 Cumulative Distribution Function (cdf) of WNWPD

The distribution function (cdf) of the wrapped lognormal distribution is given by

$$G(\theta) = \sum_{k=0}^{\infty} \frac{1}{2}\left[erf\left(\frac{\log(\theta + 2k\pi - \mu)}{\sqrt{2}\sigma} \right) - erf\left(\frac{\log(2k\pi)}{\sqrt{2}\sigma} \right) \right],$$

$$\theta, \mu \in [0, 2\pi), \theta > \mu, \sigma > 0 \text{ (Figure 2.11)} \tag{2.5.4}$$

2.5.3 Characteristic Function and Trigonometric Moments of WLND

The characteristic function (expression in closed form) of the lognormal distribution is not available in standard textbooks, such as Johnson et al. (2000) and Kendall and Stuart (1979). Therefore, an attempt is made here to derive the characteristic function of the same. Now we present numerical evaluation of the characteristic function of the lognormal distribution when $\mu = 0$, adopting the techniques suggested by Gubner (2006).

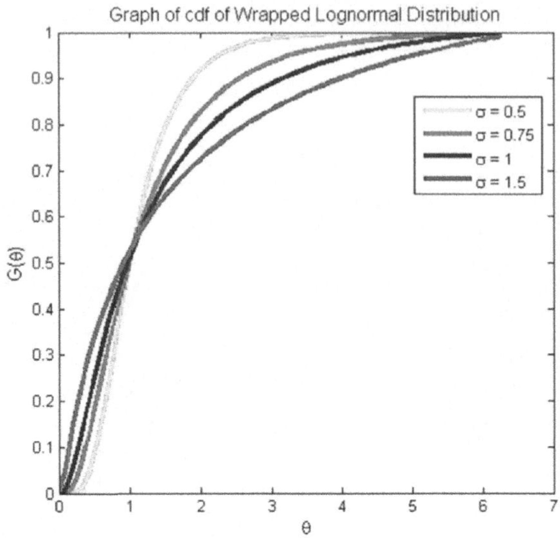

FIGURE 2.11
Graph of cdf of wrapped lognormal distribution.

The characteristic function of lognormal distribution with pdf $f(x) = \exp\left(-\{\ln x\}^2/2\sigma^2\right)$ where $x > 0$ and $\sigma > 0$ is

$$\phi_X(t) = E\left(e^{itx}\right) = \int_0^{\infty} \frac{e^{itx}}{x\sigma\sqrt{2\pi}} \exp\left(-\frac{\{\ln x\}^2}{2\sigma^2}\right) dx \text{ for } t \in \mathbb{R}$$

$$= \frac{1}{\sqrt{\pi}} \int_{-\infty}^{\infty} e^{ite^{\sqrt{2}\sigma y}} e^{-y^2} dy \text{ for } t \in \mathbb{R}$$

(making the change of variable $\frac{\ln x}{\sqrt{2}\sigma} = y$) (2.5.5)

Since the integral on the R.H.S. of Equation (2.5.5) cannot be expanded into an infinite series due to the factor exp $(it\exp(\sqrt{2}\sigma\, y))$, which is extremely oscillatory as $y \to \infty$, the characteristic function is computed for real t by adopting the methods suggested in Gubner (2006) using the n-point Hermite–Gauss quadrature. Gubner has given MATLAB® programs for generating the Hermite–Gauss weights and nodes and computing the values of the characteristic function of lognormal distribution for all real t. According to the proposition (2.2.1) from Jammalamadaka and Sen Gupta (2001), the values of the characteristic function of the wrapped lognormal distribution can be obtained by choosing an integer for t. The real and imaginary parts of the obtained values of the characteristic function of the wrapped lognormal distribution are nothing but the trigonometric moments α_p and β_p, respectively (Figure 2.12).

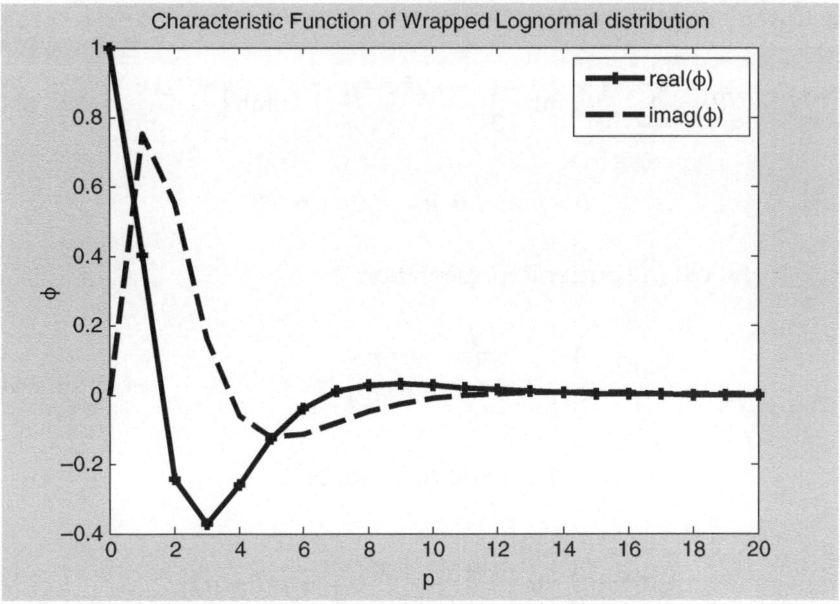

FIGURE 2.12
Graph of the characteristic function of wrapped lognormal distribution.

2.6 Other Continuous Wrapped Circular Models

Some of the other continuous wrapped circular models constructed by Dattatreya Rao et al. (2007) and Ramabhadra Sarma et al. (2009) are listed here.

2.6.1 Wrapped Logistic Distribution (WLGD)

a. Probability Density Function

$$g(\theta)=\sum_{k=-\infty}^{\infty}\frac{1}{\sigma}\left[1+\exp\left\{\frac{-(\theta+2k\pi-\mu)}{\sigma}\right\}\right]^{-2}\exp\left\{\frac{-(\theta+2k\pi-\mu)}{\sigma}\right\}, \quad (2.6.1)$$

$$\theta>\mu \text{ and } \theta,\mu\in[0,2\pi),\sigma>0$$

$$=\sum_{k=-\infty}^{\infty}\frac{1}{4\sigma}\operatorname{sech}^2\left(\frac{(\theta+2k\pi-\mu)}{2\sigma}\right),\ \theta>\mu \text{ and } \theta,\mu\in[0,2\pi),\sigma>0$$

b. Cumulative Distribution Function

$$G(\theta)=\sum_{k=-\infty}^{\infty}\frac{1}{2}\left[\tanh\left\{\frac{1}{2}\left(\frac{\theta+2k\pi-\mu}{\sigma}\right)\right\}-\tanh\left\{\frac{1}{2}\left(\frac{2k\pi}{\sigma}\right)\right\}\right] \tag{2.6.2}$$

$$\theta>\mu \text{ and } \theta,\mu\in[0,2\pi),\sigma>0$$

c. pdf and cdf in Fourier Representation

$$g(\theta)=\frac{1}{2\pi}\left[1+2\sum_{p=1}^{\infty}\frac{\pi\sigma\, p}{\sinh\pi\sigma\, p}\cos p(\theta-\mu)\right], \tag{2.6.3}$$

$$\theta>\mu \text{ and } \theta,\mu\in[0,2\pi),$$

$$G(\theta)=\frac{1}{2\pi}\left[\theta+2\sum_{p=1}^{\infty}\frac{\pi\sigma}{\sinh\pi\sigma\, p}\sin p(\theta-\mu)\right], \tag{2.6.4}$$

d. Characteristic Function

$$\begin{aligned}\varphi_W(p)&=e^{ip\mu}\frac{\pi\sigma\, p}{\sinh\pi\sigma\, p}, p\in\mathbb{Z}\\ &=\frac{\pi\sigma\, p}{\sinh\pi\sigma\, p}(\cos p\mu+i\sin p\mu),\quad p\in\mathbb{Z}\end{aligned} \tag{2.6.5}$$

e. Trigonometric Moments

$$\alpha_p=\frac{\pi\sigma\, p}{\sinh\pi\sigma\, p}\cos p\mu \tag{2.6.6}$$

$$\beta_p=\frac{\pi\sigma\, p}{\sinh\pi\sigma\, p}\sin p\mu \text{ where } p\in\mathbb{Z} \tag{2.6.7}$$

2.6.2 Wrapped Weibull Distribution (WWBD)

a. Probability Density Function

$$g(\theta)=\frac{1}{2\pi}\left[1+2\sum_{p=1}^{\infty}\{\alpha_p\cos p\theta+\beta_p\sin p\theta\}\right] \tag{2.6.8}$$

b. Cumulative Distribution Function

$$G(\theta) = \sum_{k=0}^{\infty} \left(e^{-(2k\pi)^c} - e^{-(\theta+2k\pi)^c} \right), \text{ where } \theta \in [0, 2\pi), c > 0 \tag{2.6.9}$$

c. Cumulative Distribution Function in Trigonometric form

$$G(\theta) = \frac{1}{2\pi} \left[\theta + 2 \sum_{p=1}^{\infty} \{ \alpha_p \sin p\theta - \beta_p \cos p\theta \} \right] \tag{2.6.10}$$

d. Characteristic Function

Characteristic function of wrapped Weibull distribution is not obtained in analytical form. It is evaluated by applying numerical methods and hence the trigonometric moments.

2.6.3 Wrapped Extreme-Value Distribution (WEVD)

a. Probability Density Function

$$g(\theta) = \sum_{k=-\infty}^{\infty} \frac{1}{\sigma} \exp\left(\frac{-(\theta + 2k\pi)}{\sigma} \right) \exp\left(-\exp\left(\frac{-(\theta + 2k\pi)}{\sigma} \right) \right), \tag{2.6.11}$$

for $\theta \in [0, 2\pi)$

b. Cumulative Distribution Function

$$G(\theta) = \sum_{k=-\infty}^{\infty} \left[\exp\left(-\exp\left(\frac{-(\theta + 2k\pi)}{\sigma} \right) \right) - \exp\left(-\exp\left(\frac{-(2k\pi)}{\sigma} \right) \right) \right] \tag{2.6.12}$$

c. Cumulative Distribution Function in Fourier Representation

$$G(\theta) = \frac{1}{2\pi} \left[\theta + 2 \sum_{p=1}^{\infty} \frac{1}{p} (\alpha_p \sin p\theta - \beta_p \cos p\theta) \right], \; \theta \in [0, 2\pi) \tag{2.6.13}$$

d. Characteristic Function

$$\varphi_W(p) = \Gamma(1 - ip\sigma), \text{ for } p \in \mathbb{Z} \tag{2.6.14}$$

In this case, the characteristic function involves gamma function of a complex variable. To extract trigonometric moments, it is essential to pick up the real and imaginary parts. This can be achieved using the **duplication formula**. From the tables of gamma function for complex arguments, the values of $\Gamma(z)$ can be computed. From the above computations, the real and imaginary parts of $\Gamma(1-ip\sigma)$ can be calculated, which are nothing but the trigonometric moments α_p and β_p, respectively.

2.6.4 Wrapped Binormal Distribution (WBNRD) (Dattatreya Rao et al. 2009)

a. Probability Density Function

$$g(\theta)=\begin{cases}\sum_{k=-\infty}^{\infty}\sqrt{\frac{2}{\pi}}\frac{1}{\sigma_1+\sigma_2}\exp\left[\frac{-(\theta+2k\pi-\mu)^2}{2\sigma_1^{\,2}}\right], & \theta\le\mu\\ \sum_{k=-\infty}^{\infty}\sqrt{\frac{2}{\pi}}\frac{1}{\sigma_1+\sigma_2}\exp\left[\frac{-(\theta+2k\pi-\mu)^2}{2\sigma_2^{\,2}}\right], & \theta>\mu\end{cases} \tag{2.6.15}$$

where $\theta,\mu\in[0,2\pi)$

b. Cumulative Distribution Function

$$G(\theta)=\begin{cases}\sum_{k=-\infty}^{\infty}\frac{\sigma_1}{\sigma_1+\sigma_2}\left(erf\left(\frac{(\theta+2k\pi-\mu)}{\sqrt{2}\sigma_1}\right)-erf\left(\frac{2k\pi-\mu}{\sqrt{2}\sigma_1}\right)\right), & \theta\le\mu\\ \sum_{k=-\infty}^{\infty}\frac{1}{\sigma_1+\sigma_2}\left(\begin{array}{l}-\sigma_1 erf\left(\frac{2k\pi-\mu}{\sqrt{2}\sigma_1}\right)+\sigma_1 erf\left(\frac{2k\pi}{\sqrt{2}\sigma_1}\right)\\ -\sigma_2 erf\left(\frac{2k\pi}{\sqrt{2}\sigma_2}\right)+\sigma_2 erf\left(\frac{\theta+2k\pi-\mu}{\sqrt{2}\sigma_2}\right)\end{array}\right), & \theta>\mu\end{cases} \tag{2.6.16}$$

c. pdf and cdf in Fourier Representation

$$g(\theta)=\frac{1}{2\pi}\left[1+2\sum_{p=1}^{\infty}\left(a\cos p(\theta-\mu)+b\sin p(\theta-\mu)\right)\right], \tag{2.6.17}$$

$$G(\theta)=\frac{1}{2\pi}\left[\theta+2\sum_{p=1}^{\infty}\left(\begin{array}{l}\frac{a\sin p(\theta-\mu)}{p}-\frac{b\cos p(\theta-\mu)}{p}\\ +\frac{a\sin p\mu}{p}+\frac{b\cos p\mu}{p}\end{array}\right)\right], \tag{2.6.18}$$

d. Characteristic Function

$$\varphi_W(p)=e^{ip\mu}\left[\left(\frac{\sigma_1}{\sigma_1+\sigma_2}e^{\frac{-p^2\sigma_1^2}{2}}+\frac{\sigma_2}{\sigma_1+\sigma_2}e^{\frac{-p^2\sigma_2^2}{2}}\right)\right]$$

$$+ie^{ip\mu}\left(\frac{-2}{\sqrt{\pi}}\frac{\sigma_1}{\sigma_1+\sigma_2}e^{\frac{-p^2\sigma_1^2}{2}}\sum_{n=1}^{\infty}\frac{\left(\frac{p\sigma_1}{\sqrt{2}}\right)^{2n-1}}{(2n-1)(n-1)\,!}+\frac{2}{\sqrt{\pi}}\frac{\sigma_2}{\sigma_1+\sigma_2}e^{\frac{-p^2\sigma_2^2}{2}}\sum_{n=1}^{\infty}\frac{\left(\frac{p\sigma_2}{\sqrt{2}}\right)^{2n-1}}{(2n-1)(n-1)\,!}\right)$$

for $p\in\mathbb{Z}$ (2.6.19)

$$=(a\cos p\mu-b\sin p\mu)+i(a\sin p\mu+b\cos p\mu) \tag{2.6.20}$$

where:

$$a=\left(\frac{\sigma_1}{\sigma_1+\sigma_2}e^{\frac{-p^2\sigma_1^2}{2}}+\frac{\sigma_2}{\sigma_1+\sigma_2}e^{\frac{-p^2\sigma_2^2}{2}}\right)$$

$$b=\left(\frac{-2}{\sqrt{\pi}}\frac{\sigma_1}{\sigma_1+\sigma_2}e^{\frac{-p^2\sigma_1^2}{2}}\sum_{n=1}^{\infty}\frac{\left(\frac{p\sigma_1}{\sqrt{2}}\right)^{2n-1}}{(2n-1)(n-1)\,!}+\frac{2}{\sqrt{\pi}}\frac{\sigma_2}{\sigma_1+\sigma_2}e^{\frac{-p^2\sigma_2^2}{2}}\sum_{n=1}^{\infty}\frac{\left(\frac{p\sigma_2}{\sqrt{2}}\right)^{2n-1}}{(2n-1)(n-1)\,!}\right)$$

for $p\in\mathbb{Z}$

e. Trigonometric Moments

$$\alpha_p = a\cos(p\mu) - b\sin(p\mu),\ \beta_p = a\sin(p\mu) + b\cos(p\mu) \quad \text{where } p \in \mathbb{Z}$$

2.6.5 Wrapped Half Logistic Distribution (WHLD) (Dattatreya Rao et al. 2009)

a. Probability Density Function

$$\begin{aligned} g(\theta) &= \sum_{k=-\infty}^{\infty} \frac{1}{2\sigma} \operatorname{sech}^2\left(\frac{\theta + 2k\pi - \mu}{2\sigma}\right), \theta, \mu \in [0, 2\pi) \text{ and } \theta \geq \mu, \sigma > 0 \\ &= \sum_{k=0}^{\infty} \frac{1}{2\sigma} \operatorname{sech}^2\left(\frac{\theta + 2k\pi - \mu}{2\sigma}\right) \end{aligned} \tag{2.6.21}$$

b. Cumulative Distribution Function

$$G(\theta) = \sum_{k=0}^{\infty} \tanh\left(\frac{\theta + 2k\pi - \mu}{2\sigma}\right) - \tanh\left(\frac{2k\pi}{2\sigma}\right), \theta, \mu \in [0, 2\pi) \text{ and } \theta \geq \mu \tag{2.6.22}$$

c. pdf and cdf in Fourier Representation

$$\begin{aligned} g(\theta) &= \frac{1}{2\pi}\left[1 + 2\sum_{p=1}^{\infty}\left(\alpha_p \cos p\theta + \beta_p \sin p\theta\right)\right] \\ &= \frac{1}{2\pi}\left[1 + 4\sum_{p=1}^{\infty}\left\{\begin{array}{l}\left(b_n \cos p\mu - c_n \sin p\mu\right)\cos p\theta \\ +\left(b_n \sin p\mu + c_n \cos p\mu\right)\sin p\theta\end{array}\right\}\right] \\ &= \frac{1}{2\pi}\left[1 + 4\sum_{p=1}^{\infty}\left(b_n \cos p(\theta - \mu) + c_n \sin p(\theta - \mu)\right)\right] \end{aligned}$$

$$\theta,\ \mu \in [0, 2\pi) \text{ and } \theta \geq \mu \tag{2.6.23}$$

$$\text{where } b_n = \int_0^{2\pi/p\sigma} \cos(p\sigma\, y) \frac{e^{-y}}{\left(1 + e^{-y}\right)^2} dy + \sum_{n=1}^{\infty} (-1)^{n-1} \frac{n^2 e^{-na}}{n^2 + p^2\sigma^2}$$

$$c_n = \int_0^{2\pi/p\sigma} \sin(p\sigma\, y)\frac{e^{-y}}{\left(1+e^{-y}\right)^2}dy + \sum_{n=1}^{\infty}(-1)^{n-1}\frac{np\sigma\, e^{-na}}{n^2+p^2\sigma^2},\ \text{for } p\in\mathbb{Z} \text{ and } \sigma>0$$

$$G(\theta) = \frac{1}{2\pi}\left[\theta + 4\sum_{p=1}^{\infty}\frac{1}{p}\begin{pmatrix} b_n \sin p(\theta-\mu) - c_n \cos p(\theta-\mu) \\ + b_n \sin p\mu + c_n \cos p\mu \end{pmatrix}\right] \tag{2.6.24}$$

$$\theta,\ \mu\in[0,2\pi) \text{ and } \theta \geq \mu$$

d. Characteristic Function

$$\begin{aligned}\varphi_w(p) &= 2\left(b_n \cos(p\,\mu) - c_n \sin(p\,\mu)\right) \\ &\quad + 2i\left(b_n \sin(p\,\mu) + c_n \cos(p\,\mu)\right)\end{aligned} \tag{2.6.25}$$

$$b_n = \int_0^a \cos(p\sigma\, y)\frac{e^{-y}}{\left(1+e^{-y}\right)^2}dy + \sum_{n=1}^{\infty}(-1)^{n-1}\frac{ne^{-na}}{n^2+p^2\sigma^2}\begin{pmatrix} n\cos(p\sigma\, a) \\ -p\sigma\sin(p\sigma\, a)\end{pmatrix}$$

$$c_n = \int_0^a \sin(p\sigma\, y)\frac{e^{-y}}{\left(1+e^{-y}\right)^2}dy + \sum_{n=1}^{\infty}(-1)^{n-1}\frac{ne^{-na}}{n^2+p^2\sigma^2}\begin{pmatrix} n\sin(p\sigma\, a) \\ +p\sigma\cos(p\sigma\, a)\end{pmatrix}$$

for $p\in\mathbb{Z}$

e. Trigonometric Moments

$$\alpha_p = 2\left(b_n \cos(p\,\mu) - c_n \sin(p\,\mu)\right)$$

$$\beta_p = 2\left(b_n \sin(p\,\mu) + c_n \cos(p\,\mu)\right),\ \text{for } p\in\mathbb{Z}$$

2.7 Methodology of Wrapping for Discrete Linear Models

If X is a discrete random variable on the set of integers, then reduction modulo $2\pi m(m\in\mathbb{Z}^+)$ wraps the integers on to the group of mth roots of unity, which is a subgroup of the unit circle.

$$\theta = 2\pi x(\text{mod}\, 2\pi m)$$

More precisely, θ is a mapping from a set of integers G, which is a group with respect to "+" to the set of mth roots of unity G', which is a group with respect to "." is defined as

$$\theta(x) = e^{\frac{2\pi ix}{m}} \text{ where } x \in G,\ e^{\frac{2\pi ix}{m}} \in G'$$

Then θ is called a wrapped discrete circular random variable.

Clearly, θ is a homomorphism

1. $\theta(x+y) = e^{\frac{2\pi i(x+y)}{m}} = e^{\frac{2\pi ix}{m}} e^{\frac{2\pi iy}{m}} = \theta(x)\theta(y)$

2. $\theta(0) = e^{\frac{2\pi i(0)}{m}} = e^0 = 1$ where $0 \in G, 1 \in G'$

Since θ contains a finite number of elements, they are denoted by $\theta = \{2\pi r/m\ r=0,1,2...m-1\}$ which is lattice on the unit circle.

The probability mass function, the cumulative distributive function, and the characteristic function of the wrapped discrete circular model are presented here.

2.7.1 Probability Mass Function

Suppose if θ is a wrapped discrete circular random variable, then the probability mass function is

$$Pr\left(\theta = \frac{2\pi r}{m}\right) = \sum_{k=-\infty}^{\infty} P(r+km) \text{ where } r = 0,1,2...m-1. \text{ Such that } m \in \mathbb{Z}^+ \qquad (2.7.1)$$

Further, if it is to be a circular probability mass function, it should satisfy the following properties:

1. $Pr\left(\theta = \frac{2\pi r}{m}\right) \geq 0$

2. $\sum_{r=0}^{m-1} Pr\left(\theta = \frac{2\pi r}{m}\right) = 1$

3. $Pr(\theta) = Pr(\theta + 2\pi l)$ for any integer l. Pr is periodic function.

2.7.2 Cumulative Distribution Function

Suppose if θ is a wrapped discrete circular random variable, then the distribution function of θ is denoted by $F_w(\theta)$ which is defined as

$$F_w(\theta) = \sum_{r=0}^{y} Pr(\theta = \frac{2\pi r}{m}) \qquad \text{where } y = 0,1,\ldots m-1$$

OR

$$F_w(\theta) = \sum_{k=-\infty}^{\infty} \left[F(r+km) - F(km)\right] \quad \text{where } F(X) \text{ is the distribution function of } X \tag{2.7.2}$$

2.7.3 Characteristic Function

The characteristic function of the wrapped discrete circular random variable θ is defined as

$$\begin{aligned} \varphi_\theta(p) &= \sum_{r=0}^{m-1} e^{ip\theta} Pr\left(\theta = \frac{2\pi r}{m}\right) \qquad \text{where } p \in \mathbb{Z} \\ &= \alpha_p + i\beta_p \end{aligned} \tag{2.7.3}$$

One can also see these trigonometric moments in terms of

$$\alpha_p = E(\cos p\theta), \beta_p = E(\sin p\theta) \text{where } p \in \mathbb{Z}$$

and these are related as $\rho_p = \sqrt{\alpha_p^2 + \beta_p^2}$ and $\theta = \text{Tan}^{-1}\left(\frac{y}{x}\right)$.

2.7.4 Central Trigonometric Moments

The pth central trigonometric moment of θ is defined as

$$\begin{aligned} \varphi_\theta^*(p) &= E\left[e^{ip(\theta-\mu)}\right] \\ &= \alpha_p^* + i\beta_p^* \end{aligned}$$

Now the skewness for the circular distribution is defined as $\gamma_1 = \beta_2^* / (V_o)^{\frac{3}{2}}$ and kurtosis for the circular distribution is defined as $\gamma_2 = \alpha_2^* - (1 - V_o)^4 / (V_o)^2$.

2.8 Wrapped Binomial Distribution (WBD)

The binomial distribution with parameters n, p_1 ($q_1 = 1 - p_1$) is defined as the distribution of a random variable X (Girija et al. 2014a), then for $x = 0,1,2 \ldots n$ and $p_1 + q_1 = 1$ where $n \in \mathbb{Z}^+, 0 < p_1 < 1$.

The probability mass function is

$$f(x) = \Pr(X = x) = n_{c_x} p_1{}^x q_1{}^{n-x}$$

which is probability of number of success in n trails.

The cumulative distribution function is

$$F(x) = I_{q_1}(n - x, x + 1)$$

where I_{q_1} is the regularized incomplete Beta function which is defined as

$$I_{q_1}(a,b) = \frac{B(q_1; a, b)}{B(a,b)}$$

where $B(q_1; a, b)$ is the incomplete Beta function and $B(a,b)$ is the Beta function.

The characteristic function is

$$\phi_X(t) = (q_1 + p_1 e^{it})^n, \quad t \in \mathbb{R}$$

The probability mass function, cumulative distribution function, and characteristic function of wrapped binomial distribution are defined here.

2.8.1 Probability Mass Function of Wrapped Binomial Distribution

Suppose if X follows binomial distribution, then the probability mass function of the wrapped binomial distribution is defined by using Equation (2.7.1)

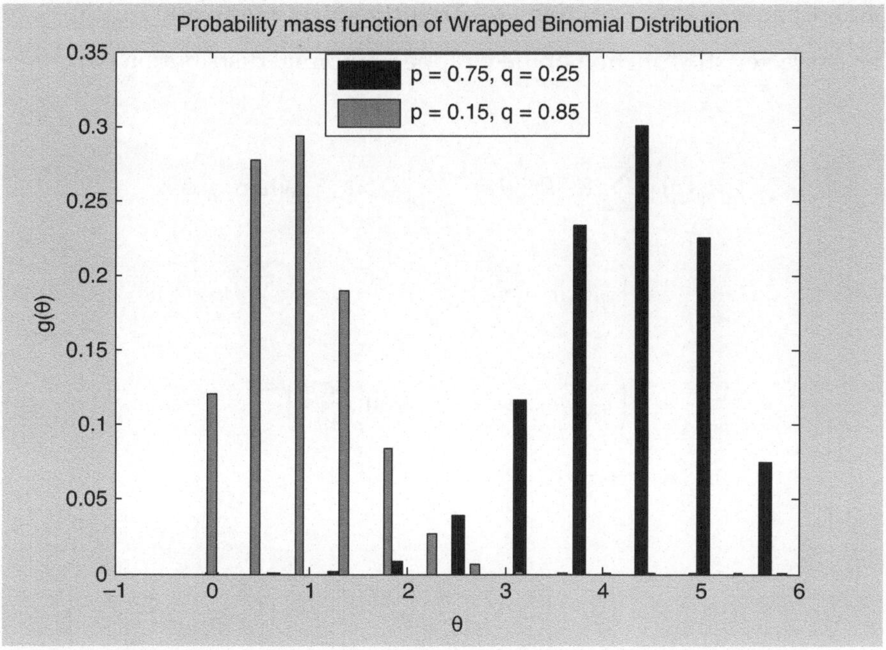

FIGURE 2.13
Probability mass function of wrapped binomial distribution.

$$Pr\left(\theta = \frac{2\pi r}{m}\right) == \sum_{k=0}^{\left[\frac{n-r}{m}\right]} n_{C_{r+km}} {p_1}^{r+km} {q_1}^{n-r-km} \text{ where } n \geq m-1 \ni m, n \in \mathbb{Z}^+$$

Here p_1 is the probability of success, and q_1 is the probability of failure. Where n, p_1 are the parameters (Figure 2.13).

2.8.2 Cumulative Distribution Function of Wrapped Binomial Distribution

The cumulative distribution function of the Wrapped binomial distribution is

$$F_w(\theta) = \sum_{r=0}^{y} \left[\sum_{k=0}^{\left[\frac{n-r}{m}\right]} n_{C_{r+km}} {p_1}^{r+km} {q_1}^{n-r-km} \right] \qquad \text{where } y = 0,1,\ldots m-1$$

2.8.3 Characteristic Function and Trigonometric Moments

The characteristic function of the wrapped binomial distribution is

$$\varphi_\theta(p) = \sum_{r=0}^{m-1} e^{ip\theta} Pr\left(\theta = \frac{2\pi r}{m}\right) = \rho_p e^{i\mu_p} \quad \text{where } p \in \mathbb{Z}$$

$$= \left(q_1 + p_1\, e^{\frac{2\pi i p}{m}}\right)^n$$

$$= \left(q_1 + p_1 \cos\frac{2\pi p}{m} + i p_1 \sin\frac{2\pi p}{m}\right)^n$$

$$= \alpha_p + i\beta_p$$

$$\text{Then } \alpha_p = \left(q_1^2 + p_1^2 + 2q_1 p_1 \cos\frac{2\pi p}{m}\right)^{\frac{n}{2}} \cos n\left[\operatorname{Tan}^{-1}\left(\frac{p_1 \sin\frac{2\pi p}{m}}{q_1 + p_1 \cos\frac{2\pi p}{m}}\right)\right]$$

$$\text{and } \beta_p = \left(q_1^2 + p_1^2 + 2q_1 p_1 \cos\frac{2\pi p}{m}\right)^{\frac{n}{2}} \sin n\left[\operatorname{Tan}^{-1}\left(\frac{p_1 \sin\frac{2\pi p}{m}}{q_1 + p_1 \cos\frac{2\pi p}{m}}\right)\right]$$

Here α_p, β_p are called pth trigonometric moments.

Clearly $\rho_p = \sqrt{\alpha_p^2 + \beta_p^2}$

$$\rho_p = \sqrt{\left(q_1^2 + p_1^2 + 2q_1 p_1 \cos\frac{2\pi p}{m}\right)^n}$$

And $\mu_p = \operatorname{Tan}^{-1}\left[\frac{\beta_p}{\alpha_p}\right]$

$$\therefore \mu_p = n\,\mathrm{Tan}^{-1}\left(\frac{p_1 \sin\frac{2\pi p}{m}}{q_1 + p_1 \cos\frac{2\pi p}{m}}\right)$$

Now the circular mean direction, $\mu_1 = n\,\mathrm{Tan}^{-1}\left(\frac{p_1 \sin\frac{2\pi}{m}}{q_1 + p_1 \cos\frac{2\pi}{m}}\right)$

Now ρ_1 represents the concentration toward the mean direction,

$$\rho_1 = \sqrt{\left(q_1^2 + p_1^2 + 2q_1p_1 \cos\frac{2\pi}{m}\right)^n}$$

In general $\mu_1 = \mu$ and $\rho_1 = \rho$.

Now the circular variance, $V_o = 1 - \rho$

$$V_o = 1 - \sqrt{\left(q_1^2 + p_1^2 + 2q_1p_1 \cos\frac{2\pi}{m}\right)^n}$$

The circular standard deviation

$$\sigma_o = \sqrt{-2\log(1 - V_o)}$$

$$\sigma_o = \sqrt{\log\left(\frac{1}{\left(q_1^2 + p_1^2 + 2q_1p_1 \cos\frac{2\pi}{m}\right)^n}\right)}$$

Central trigonometric moments

The pth central trigonometric moment of θ is denoted by $\varphi_\theta^*(p)$, and it is defined as

$$\begin{aligned}\varphi_\theta^*(p) &= E\left[e^{ip(\theta-\mu)}\right] = \alpha_p^* + i\beta_p^* = E\left[e^{ip\theta}e^{-ip\mu}\right] \\ &= e^{-ip\mu}E\left[e^{ip\theta}\right] \\ &= \left(\alpha_p \cos p\mu + \beta_p \sin p\mu\right) + i\left(\beta_p \cos p\mu - \alpha_p \sin p\mu\right)\end{aligned}$$

where $\alpha_p^* = \alpha_p \cos p\mu + \beta_p \sin p\mu$ and $\beta_p^* = \beta_p \cos p\mu - \alpha_p \sin p\mu$.

The skewness is

$$\begin{aligned}\gamma_1 &= \frac{\beta_2^*}{(V_o)^{\frac{3}{2}}} \\ &= \frac{\beta_2 \cos 2\mu - \alpha_2 \sin 2\mu}{\left[1 - \sqrt{\left(q_1^2 + p_1^2 + 2q_1p_1 \cos \frac{2\pi}{m}\right)^n}\right]^{\frac{3}{2}}}\end{aligned}$$

The kurtosis is

$$\begin{aligned}\gamma_2 &= \frac{\alpha_2^* - (1 - V_o)^4}{(V_o)^2} \\ &= \frac{\left(\alpha_2 \cos 2\mu + \beta_2 \sin 2\mu\right) - \left[\sqrt{\left(q_1^2 + p_1^2 + 2q_1p_1 \cos \frac{2\pi}{m}\right)^n}\right]^4}{\left[1 - \sqrt{\left(q_1^2 + p_1^2 + 2q_1p_1 \cos \frac{2\pi}{m}\right)^n}\right]^2} \\ &= \frac{\left(\alpha_2 \cos 2\mu + \beta_2 \sin 2\mu\right) - \left[q_1^2 + p_1^2 + 2q_1p_1 \cos \frac{2\pi}{m}\right]^{2n}}{\left[1 - \sqrt{\left(q_1^2 + p_1^2 + 2q_1p_1 \cos \frac{2\pi}{m}\right)^n}\right]^2}\end{aligned}$$

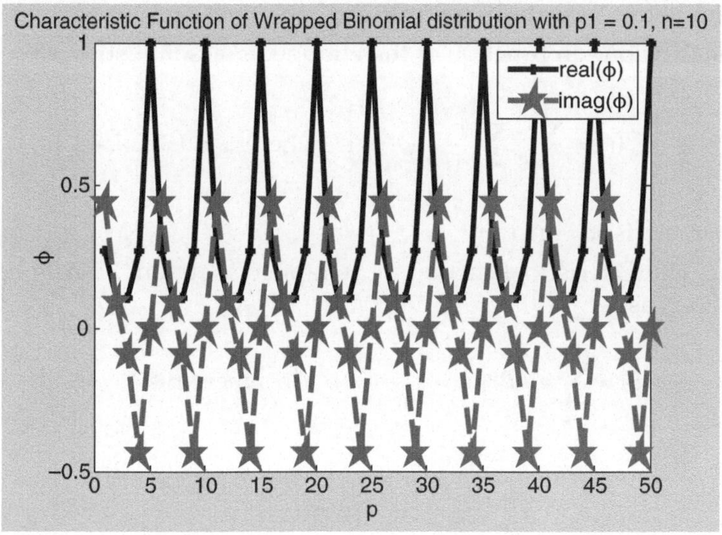

FIGURE 2.14
Characteristic function of wrapped binomial distribution.

2.8.4 Population Characteristics of Wrapped Binomial Distribution

On the lines of expressions available in Mardia and Jupp (2000), the population characteristics for the wrapped binomial distribution are computed. For computation of characteristics, some arbitrary values are taken for the parameter p_1 (Figure 2.14).

2.9 Other Discrete Wrapped Circular Models

Some of the other discrete wrapped circular models are listed here.

2.9.1 Wrapped Poisson Distribution (WPD)

The probability mass function, cumulative distribution function, and characteristic function of the wrapped Poisson distribution (Girija et al. 2014b) are defined here.

a. Probability Mass Function

   Suppose if X follows the Poisson distribution with parameter $\lambda > 0$, then the probability mass function of the wrapped Poisson distribution is

$$Pr\left(\theta = \frac{2\pi r}{m}\right) = \sum_{k=0}^{\infty} \frac{e^{-\lambda}\lambda^{r+km}}{(r+km)!}$$

b. Cumulative Distribution Function

The distribution function of the wrapped Poisson distribution is

$$F_w(\theta) = \sum_{r=0}^{y}\left[\sum_{k=0}^{\infty}\frac{e^{-\lambda}\lambda^{r+km}}{(r+km)!}\right] \quad \text{where } y = 0,1,\ldots m-1$$

c. Characteristic Function

The characteristic function of wrapped Poisson distribution is defined as

$$\varphi_\theta(p) = \sum_{r=0}^{m-1} e^{ip\theta} Pr\left(\theta = \frac{2\pi r}{m}\right) = \rho_p e^{i\mu_p} \quad \text{where } p \in \mathbb{Z}$$

$$= e^{-\lambda\left[1-\cos\frac{2\pi p}{m}\right]}\left[\cos\left(\lambda\sin\frac{2\pi p}{m}\right) + i\sin\left(\lambda\sin\frac{2\pi p}{m}\right)\right]$$

$$= \alpha_p + i\beta_p$$

where $\alpha_p = e^{-\lambda(1-\cos\frac{2\pi p}{m})}\cos(\lambda\sin\frac{2\pi p}{m})$ and $\beta_p = e^{-\lambda\left(1-\cos\frac{2\pi p}{m}\right)}\sin(\lambda\sin\frac{2\pi p}{m})$.

Here α_p, β_p are called pth trigonometric moments.

d. Population Characteristics

$$\rho_p = \sqrt{\alpha_p^2 + \beta_p^2} = e^{-\lambda\left(1-\cos\frac{2\pi p}{m}\right)}$$

$$\mu_p = \tan^{-1}\left(\frac{\beta_p}{\alpha_p}\right) = \tan^{-1}\left[\tan\left(\lambda\sin\frac{2\pi p}{m}\right)\right] = \lambda\sin\frac{2\pi p}{m}$$

The circular mean direction is $\mu_1 = \lambda\sin\frac{2\pi}{m}$ and

Concentration is $\rho_1 = e^{-\lambda\left(1-\cos\frac{2\pi}{m}\right)}$

In general $\mu_1=\mu$ and $\rho_1=\rho$

The circular variance is $V_o = 1-\rho = 1-e^{-\lambda\left(1-\cos\frac{2\pi}{m}\right)}$

The circular standard deviation is

$$\sigma_o = \sqrt{-2\log(1-V_o)} = \sqrt{-2\log\left(e^{-\lambda\left(1-\cos\frac{2\pi}{m}\right)}\right)}$$

e. Central Trigonometric Moments

The pth central trigonometric moment of θ is defined as

$$\varphi_\theta^*(p) = E\left[e^{ip(\theta-\mu)}\right]$$

$$= e^{-\lambda\left(1-\cos\frac{2\pi p}{m}\right)}\left[\cos(\mu_p - p\mu) + i\sin(\mu_p - p\mu)\right] = \alpha_p^* + i\,\beta_p^*$$

where $\alpha_p^* = e^{-\lambda\left(1-\cos\frac{2\pi p}{m}\right)}\cos(\mu_p - p\mu)$ and $\beta_p^* = e^{-\lambda\left(1-\cos\frac{2\pi p}{m}\right)}\sin(\mu_p - p\mu)$.

Skewness is

$$\gamma_1 = \frac{\beta_2^*}{(V_o)^{\frac{3}{2}}} = \frac{e^{-\lambda\left(1-\cos\frac{4\pi}{m}\right)}\sin(\mu_2 - 2\mu)}{\left[1-e^{-\lambda\left(1-\cos\frac{2\pi}{m}\right)}\right]^{\frac{3}{2}}}$$

Kurtosis is

$$\gamma_2 = \frac{\alpha_2^* - (1-V_o)^4}{(V_o)^2}$$

$$= \frac{e^{-\lambda\left(1-\cos\frac{4\pi}{m}\right)}\cos(\mu_2 - 2\mu) - e^{-4\lambda\left(1-\cos\frac{2\pi}{m}\right)}}{\left[1-e^{-\lambda\left(1-\cos\frac{2\pi}{m}\right)}\right]^2}$$

2.9.2 Wrapped Logarithmic Distribution (WLGRD) (Srihari et al. 2018)

a. Probability Mass Function

Suppose if X follows logarithmic distribution with the parameter $p_1 (0 < p_1 < 1)$, then the probability mass function of the wrapped logarithmic distribution (WLGRD) is

$$Pr\left(\theta = \frac{2\pi r}{m}\right) = \sum_{k=0}^{\infty} \frac{-1}{\ln(1-p_1)} \frac{p_1^{r+km}}{r+km}$$

where $r = 1, 2, \ldots m$ such that $m \in \mathbb{Z}^+$

b. Cumulative Distribution Function

The distribution function of the wrapped logarithmic distribution (WLGRD) is

$$F_w(\theta) = \frac{1}{\ln(1-p_1)} \begin{bmatrix} B(p_1; r+1,0) - B(p_1;1,0) \\ \quad + B(p_1; r+m+1,0) - B(p_1; m+1,0) \\ \quad + B(p_1; r+2m+1,0) - B(p_1; 2m+1,0) + \dots \end{bmatrix}$$

$$= \frac{1}{\ln(1-p_1)} \left[\sum_{k=0}^{\infty} B(p_1; r+1+km,0) - \sum_{k=0}^{\infty} B(p_1; km+1,0) \right]$$

where B is the incomplete Beta function.

c. Characteristic Function and Trigonometric Moments

The characteristic function of the wrapped logarithmic distribution is

$$\varphi_\theta(p) = \sum_{r=1}^{m} e^{ip\theta} Pr(\theta = \frac{2\pi r}{m}) \quad \text{where } p \in \mathbb{Z}$$

$$= \frac{-1}{\ln(1-p_1)} \sum_{r=1}^{m} e^{ip\theta} \left[\sum_{k=0}^{\infty} \frac{p_1^{\,r+km}}{r+km} \right]$$

where $x = 1 - p_1 \cos\frac{2\pi p}{m}$ and $y = p_1 \sin\frac{2\pi p}{m}$.

Then $x^2 + y^2 = R^2$ and $\theta = \tan^{-1}\left(\frac{y}{x}\right)$

$$\frac{\ln(x-iy)}{\ln(1-p_1)} = \frac{\ln(R\cos\theta - iR\sin\theta)}{\ln(1-p_1)}$$

$$= \frac{\ln\left(Re^{-i\theta}\right)}{\ln(1-p_1)} = \frac{\ln R - i\theta}{\ln(1-p_1)}$$

$$= \frac{\ln R}{\ln(1-p_1)} - i\frac{\theta}{\ln(1-p_1)}$$

$$= \alpha_p + i\beta_p$$

where $\alpha_p = \frac{\ln R}{\ln(1-p_1)}$, $\beta_p = \frac{-\theta}{\ln(1-p_1)}$

$$R=\sqrt{x^2+y^2}$$
$$=\sqrt{1+p_1^2\cos^2\frac{2\pi p}{m}-2p_1\cos\frac{2\pi p}{m}+p_1^2\sin^2\frac{2\pi p}{m}}$$
$$=\sqrt{1+p_1^2-2p_1\cos\frac{2\pi p}{m}}$$

$$\theta=\tan^{-1}\left(\frac{y}{x}\right)=\tan^{-1}\left(\frac{p_1\sin\frac{2\pi p}{m}}{1-p_1\cos\frac{2\pi p}{m}}\right)$$

Then $\alpha_p=\dfrac{\ln\left(\sqrt{1+p_1^2-2p_1\cos\frac{2\pi p}{m}}\right)}{\ln(1-p_1)}$, $\beta_p=\dfrac{-\tan^{-1}\left(\dfrac{p_1\sin\frac{2\pi p}{m}}{1-p_1\cos\frac{2\pi p}{m}}\right)}{\ln(1-p_1)}$

d. Population Characteristics

Now $\rho_p=\sqrt{\alpha_p^2+\beta_p^2}$

$$\rho_p=\sqrt{\frac{\left[\ln\left(\sqrt{1+p_1^2-2p_1\cos\frac{2\pi p}{m}}\right)\right]^2+\left[\tan^{-1}\left(\dfrac{p_1\sin\frac{2\pi p}{m}}{1-p_1\cos\frac{2\pi p}{m}}\right)\right]^2}{\left[\ln(1-p_1)\right]^2}}$$

and $\mu_p=\tan^{-1}\left(\dfrac{\beta_p}{\alpha_p}\right)$

$$\mu_p=\tan^{-1}\left(\frac{-\tan^{-1}\left(\dfrac{p_1\sin\frac{2\pi p}{m}}{1-p_1\cos\frac{2\pi p}{m}}\right)}{\ln\left(\sqrt{1+p_1^2-2p_1\cos\frac{2\pi p}{m}}\right)}\right)$$

The circular mean direction is $\mu_1 = \tan^{-1}\left(\dfrac{-\tan^{-1}\left(\dfrac{p_1 \sin\frac{2\pi}{m}}{1-p_1\cos\frac{2\pi}{m}}\right)}{\ln\left(\sqrt{1+p_1^2-2p_1\cos\frac{2\pi}{m}}\right)}\right)$

The concentration is

$$\rho_1 = \sqrt{\frac{\left[\ln\left(\sqrt{1+p_1^2-2p_1\cos\frac{2\pi}{m}}\right)\right]^2+\left[\tan^{-1}\left(\dfrac{p_1\sin\frac{2\pi}{m}}{1-p_1\cos\frac{2\pi}{m}}\right)\right]^2}{\left[\ln(1-p_1)\right]^2}}$$

In general $\mu_1 = \mu$ and $\rho_1 = \rho$

The circular variance is $V_o = 1-\rho$

$$V_o = 1-\sqrt{\frac{\left[\ln\left(\sqrt{1+p_1^2-2p_1\cos\frac{2\pi}{m}}\right)\right]^2+\left[\tan^{-1}\left(\dfrac{p_1\sin\frac{2\pi}{m}}{1-p_1\cos\frac{2\pi}{m}}\right)\right]^2}{\left[\ln(1-p_1\right]^2}}$$

The circular standard deviation is

$$\sigma_o = \sqrt{-2\log(1-V_o)} = \sqrt{-2\log\rho}$$

$$= \sqrt{-2\log\left[\sqrt{\frac{\left[\ln\left(\sqrt{1+p_1^2-2p_1\cos\frac{2\pi}{m}}\right)\right]^2+\left[\tan^{-1}\left(\dfrac{p_1\sin\frac{2\pi}{m}}{1-p_1\cos\frac{2\pi}{m}}\right)\right]^2}{\left[\ln(1-p_1)\right]^2}}\right]}$$

e. Central Trigonometric Moments

The pth central trigonometric moments are

$$\varphi_\theta^*(p) = E\left[e^{ip(\theta-\mu)}\right] = \alpha_p^* + i\beta_p^*$$

$$\alpha_p^* = \alpha_p \cos p\mu + \beta_p \sin p\mu \text{ and}$$

$$\beta_p^* = \beta_p \cos p\mu - \alpha_p \sin p\mu$$

The skewness is

$$\gamma_1 = \frac{\beta_2^*}{(V_o)^{\frac{3}{2}}}$$

$$= \frac{\beta_2 \cos 2\mu - \alpha_2 \sin 2\mu}{\left[1-\sqrt{\frac{\left[\ln\left(\sqrt{1+p_1^2-2p_1\cos\frac{2\pi}{m}}\right)\right]^2 + \left[\tan^{-1}\left(\frac{p_1 \sin\frac{2\pi}{m}}{1-p_1\cos\frac{2\pi}{m}}\right)\right]^2}{\left[\ln(1-p_1)\right]^2}}\right]^{\frac{3}{2}}}$$

The kurtosis is

$$\gamma_2 = \frac{\alpha_2^* - (1-V_o)^4}{(V_o)^2}$$

$$= \frac{(\alpha_2 \cos 2\mu + \beta_2 \sin 2\mu) - \left[\sqrt{\frac{\left[\ln\left(\sqrt{1+p_1^2-2p_1\cos\frac{2\pi}{m}}\right)\right]^2 + \left[\tan^{-1}\left(\frac{p_1 \sin\frac{2\pi}{m}}{1-p_1\cos\frac{2\pi}{m}}\right)\right]^2}{\left[\ln(1-p_1)\right]^2}}\right]^4}{\left[1-\sqrt{\frac{\left[\ln\left(\sqrt{1+p_1^2-2p_1\cos\frac{2\pi}{m}}\right)\right]^2 + \left[\tan^{-1}\left(\frac{p_1 \sin\frac{2\pi}{m}}{1-p_1\cos\frac{2\pi}{m}}\right)\right]^2}{\left[\ln(1-p_1)\right]^2}}\right]^2}$$

References

Akinsete, A., Famoye, F. and Lee, C. 2008. The Beta-Pareto distribution, *Statistics*, 42(6), 547–563.

Alzaatreh, A., Famoye, F. and Lee, C. 2012. Gamma-Pareto Distribution and Its Applications, *Journal of Modern Applied Statistical Methods*, 11(1), 78–94.

Alzaatreh, A., Famoye, F. and Lee, C. 2013. Weibull-Pareto Distribution and Its Applications. *Communication in Statistics-Theory and Methods*, 42, 1673–1691.

Carslaw, H.S. 1930. *Introduction to the Theory of Fourier's Series and Integrals*, 3rd ed., Dover, New York.

Dattatreya Rao, A.V., Ramabhadra Sarma, I. and Girija, S.V.S. 2007. On Wrapped Version of Some Life Testing Models, *Communication in Statistics: Theory and Methods*, 36(11), 2027–2035.

Dattatreya Rao, A.V., Ramabhadra Sarma, I. and Girija, S.V.S. 2009. On Characteristic Functions of Wrapped Half Logistic and Binormal Distributions, *International Journal of Statistics and Systems*, 4(1), 33–45.

Flaih, A.H., Elsalloukh, E.M. and Milanova, M. 2012. The Exponentiated Inverted Weibull Distribution, *Applied Mathematics and Information Sciences*, 6(2), 167–171.

Girija, S.V.S. 2010. *Construction of New Circular Models*, VDM VERLAG, Germany.

Girija, S.V.S., Dattatreya Rao, A.V. and Srihari, V.L.N. 2014a. On Wrapped Binomial Model Characteristics, *Mathematics and Statistics*, 2(7), 231–234.

Girija, S.V.S., Dattatreya Rao, A.V. and Srihari, V.L.N. 2014b. On Characteristic Function of Wrapped Poisson Distribution, *International Journal of Mathematical Archive*, 5(5), 168–173.

Gubner, J.A. 2006. A New Formula for Lognormal Characteristic Functions, *Vehicular Technology, IEEE Transactions*, 55(5), 1668–1671.

Gupta, R.D. and Kundu, D. 2001. Exponentiated Exponential Family: An Alternative to Gamma and Weibull Distributions, *Biometrical Journal*, 43, 117–130.

Jammalamadaka, S.R. and Sen Gupta, A. 2001. *Topics in Circular Statistics*, World Scientific Press, Singapore.

Johnson, N.L., Samuel, K. and Balakrishnan, N. 2000. Continuous Univariate Distributions Vol. 1 and 2, *Wiley Series in Probability and Statistics*, Wiley, New York.

Kareema, A.K. and Boshi, M.A. 2013. Exponential Pareto Distribution, *Mathematical Theory and Modeling*, 3(5), 135–146.

Kendall, M. and Stuart, A. 1979. *The Advanced Theory of Statistics*, 4th ed., vol. 2. Griffin, London, UK.

Mardia, K.V. 1972. *Statistics of Directional Data*, Academic Press, London, UK.

Mardia, K.V. and Jupp, P.E. 2000. *Directional Statistics*, John Wiley & Sons, Chichester, UK.

Merovci, F. and Puka, L. 2014. Transmuted Pareto Distribution, *Probstat Forum*, 7, 1–11.

Mudholkar, G.S., Srivastava, D.K. and Freimer, M. 1995. The Exponentiated Weibull Family: A reanalysis of the Bus-Motor Failure Data, *Technometrics*, 37, 436–445.

Nadarajah, S. and Kotz, S. 2006. The Exponentiated Type Distributions, *Acta Applicandae Mathematica*, 92(2), 97–111.

Pickands, J. 1975. Statistical Inference Using Extreme Order Statistics, *Annals of Statistics* 3, 119–131.

Ramabhadra Sarma, I., Dattatreya Rao, A.V. and Girija, S.V.S. 2009. On Characteristic Functions of Wrapped Half Logistic and Binormal Distributions, *International Journal of Statistics and Systems*, 4(1), 33–45.

Rao, C.R. and Mitra, S.K. 1975. *Formulae and Tables for Statistical Work*, Statistical Publishing Society, Calcutta, India.

Silva, R.B., Barreto-Souza, W. and Cordeiro, G. M. 2010. A New Distribution with Decreasing, Increasing and Upside-Down Bathtub Failure Rate, *Computational Statistics and Data Analysis* 54, 935–944.

Srihari, G.V.L.N., Girija, S.V.S. and Dattatreya Rao, A.V. 2018. On Characteristics of Wrapped Logarithmic Model, *International Journal of Current Trends in Science and Technology*, 8(4), 20413–20419.

Subrahmanyam, P.S., Dattatreya Rao, A.V. and Girija, S.V.S. 2017a. On Wrapping of Exponentiated Inverted Weibull Distribution, *International Journal for Innovative Research in Science & Technology*, 3(11), 18–25.

Subrahmanyam, P.S., Dattatreya Rao, A.V. and Girija, S.V.S. 2017b. On Wrapping of New Weibull Pareto Distribution, *International Journal of Advanced Research and Review*, 2(4), 10–20.

Suleman, N., Albert, L. 2015. The New Weibull-Pareto Distribution, *Pakistan Journal of Statistics and Operation Research* XI(1), 103–114.

Tahir, M.H., Gauss, M.C., Ayman, A., Mansoor, M. and Zubair, M. 2014. A New Weibull-Pareto Distribution: Properties and Applications, *Communications in Statistics: Simulation and Computation* 45(10), 3548–3567.

3

Stereographic Circular Models

3.1 Introduction

Smooth and bijective mapping called stereographic projection is a mapping that projects a sphere onto a plane. The projection is defined on the entire sphere, except at the projection point. Its 2D version can be seen as the mapping that projects the entire circle onto a line.

Minh and Farnum (2003) developed a new method of generating probability distributions, applying a planar stereographic projection. Taking this as a cue, Toshihiro et al. (2013) constructed symmetric unimodal circular distributions applying inverse stereographic projection but referred them as stereographic distributions. On these lines, Dattatreya Rao et al. (2013) discussed Cauchy-type distributions.

In view of the importance of lognormal, logistic, and double-Weibull distributions in various fields of applications, it is considered to construct stereographic circular distributions projecting them onto the entire circle and listed as stereographic logistic distribution (SLGD), stereographic double-Weibull distribution (SDWD), stereographic lognormal distribution (SLND), stereographic extreme-value distribution (SEVD), and stereographic reflected gamma distribution (SRGD).

Having constructed the new stereographic circular models, graphs of pdf, cdf, and characteristic functions of the above distributions are depicted mentioning the nature of distributions based on population characteristics in the subsequent sections of this chapter.

3.2 Methodology of Inverse Stereographic Projection

Probability distributions (both circular and linear) can be generated by applying Stereographic Projection (Phani et al. 2012a), which yields a one-to-one correspondence between the points on the unit circle and those on the real line. Inverse stereographic projection is defined by a one-to-one mapping given by $T(\theta)=x=u+v\tan(\frac{\theta}{2})$, where $x\in(-\infty,\infty)$, $\theta\in(-\pi,\pi)$, $u\in\mathbb{R}$, and $v>0$. Suppose x is randomly chosen in the interval $(-\infty,\infty)$. Let $F(x)$ and $f(x)$ denote the cumulative distribution and the probability density functions of the random variable X, respectively. Then $T^{-1}(x)=\theta=2\tan^{-1}\left\{\frac{(x-u)}{v}\right\}$ is a random point on the unit circle. Let $G(\theta)$ and $g(\theta)$ denote the cumulative distribution and the probability density functions of this random point θ, respectively. Then $G(\theta)$ and $g(\theta)$ can be derived in terms of $F(x)$ and $f(x)$ using the following lemma.

Lemma 3.2.1

If $v>0$, $T^{-1}(x)=\theta=2\tan^{-1}\left(\frac{x-u}{v}\right)$ increases monotonically from $-\pi$ to π as x increases from $-\infty$ to ∞.

Proof: Suppose $v>0$,

$$T^{-1}(x)=\theta=2\tan^{-1}\left(\frac{x-u}{v}\right)$$

$$\frac{d\theta}{dx}=\frac{2v}{v^2+(x-u)^2}>0,\ \forall\ x\in(-\infty,\infty),\text{ since } v>0$$

$$\frac{d\theta}{dx}>0,\forall\, x\in(-\infty,\infty)$$

Therefore, $T^{-1}(x)=\theta=2\tan^{-1}\left(\frac{x-u}{v}\right)$ increases monotonically from $-\pi$ to π as x increases from $-\infty$ to ∞.

Theorem 3.2.1

For $v>0$,

1. $$G(\theta)=F\left(u+v\tan\left(\frac{\theta}{2}\right)\right)=F(x(\theta)) \tag{3.2.1}$$

2. $$g(\theta)=v\left(\frac{1+\tan^2\left(\frac{\theta}{2}\right)}{2}\right)f\left(u+v\tan\left(\frac{\theta}{2}\right)\right) \tag{3.2.2}$$

Proof: Suppose $v > 0$, for any complex number $C = u - iv$

$$T^{-1}(x) = \theta = 2\tan^{-1}\left(\frac{x-u}{v}\right) \text{ or } x = T(\theta) = u + v\tan\left(\frac{\theta}{2}\right)$$

1. The cumulative distribution function is given by

$$G(\theta) = \Pr\left(T^{-1}(x) \le \theta\right)$$

$$= \Pr\left(x \le T(\theta)\right) \text{ (from Lemma 3.2.1)}$$

$$G(\theta) = F(x) = F\left(u + v\tan\left(\frac{\theta}{2}\right)\right)$$

2. The probability density function is given by

$$g(\theta) = \frac{d\left(G(\theta)\right)}{d\theta}$$

$$= \frac{d\left(F\left(u + v\tan\left(\frac{\theta}{2}\right)\right)\right)}{d\theta}$$

$$g(\theta) = \frac{v}{2}\sec^2\left(\frac{\theta}{2}\right) f\left(u + v\tan\left(\frac{\theta}{2}\right)\right)$$

With $u = 0$ and $v = 1$ the inverse stereographic projection is modified and employed hereafter for construction of stereographic circular models. By inducing modified inverse stereographic projection, the probability density function of stereographic circular model is

$$g(\theta) = \frac{1}{2}\sec^2\left(\frac{\theta}{2}\right) f\left(\tan\left(\frac{\theta}{2}\right)\right), \; -\pi < \theta < \pi$$

3.3 The Characteristic Function of a Stereographic Circular Model

The characteristic function of a circular model (Phani 2013) with the probability density function $g(\theta)$ is defined as $\varphi_p(\theta) = \int_0^{2\pi} e^{ip\theta} g(\theta)\, d\theta \;, p \in \mathbb{Z}$. Ramabhadra Sarma et al. (2009, 2011) derived the characteristic functions of some new wrapped models based on the proposition (Jammalamadaka and Sen Gupta 2001, p. 31). This proposition cannot be applied directly in case of a stereographic circular model. However, the characteristic function of a stereographic circular model can be obtained in terms of the respective linear model (Lukacs 1970) using the following theorem.

Theorem 3.3.1

Let X be a random variable with distribution function $F(x)$ and suppose that $S(x)$ is a finite and single-valued function of x. The characteristic function of $\varphi_Y(t)$ of the random variable $Y = S(x)$ is then given by

$$\varphi_Y(t) = E\left(e^{itY}\right) = E\left(e^{itS(X)}\right) = \int_{-\infty}^{\infty} e^{itS(X)} dF(x).$$

By applying the above theorem, the characteristic function of a stereographic circular model is proposed in Theorem 3.3.2.

Theorem 3.3.2

If $G(\theta)$ and $g(\theta)$ are the cdf and the pdf of the stereographic circular model, and $F(x)$ and $f(x)$ are the cdf and the pdf of the respective linear model, then the characteristic function of a stereographic circular model is $\varphi_{X_S}(p) = \varphi_{2\tan^{-1}\left(\frac{x}{v}\right)}(p), p \in \mathbb{Z}$.

Proof: For $p \in \mathbb{Z}$, $\varphi_{X_S}(p) = \int_{-\pi}^{\pi} e^{ip\theta} d(G(\theta))$

$$= \int_{-\pi}^{\pi} e^{ip\theta} d\left(F\left(v\tan\frac{\theta}{2}\right)\right) = \int_{-\infty}^{\infty} e^{ip\left(2\tan^{-1}\left(\frac{x}{v}\right)\right)} f(x)dx,$$

taking $x = v\tan\left(\dfrac{\theta}{2}\right)$

$$= \varphi_{2\tan^{-1}\left(\frac{x}{v}\right)}(p)$$

As the integral is not tractable analytically, it is evaluated numerically using Weddle's rule, Gauss–Laguerre and Gauss–Hermite methods (Abramowitz and Stegun 1965), hence, plotting the characteristic function.

3.4 Stereographic Logistic Distribution (SLGD)

A random variable X on the real line is said to have the logistic distribution (Dattatreya Rao et al. 2016) with location parameter γ and scale parameter $\lambda>0$, if the probability density function and the cumulative distribution function of X for $x,\gamma \in \mathbb{R}$ and $\lambda > 0$ are given by

$$\begin{aligned} f(x) &= \frac{1}{\lambda}\left[1+\exp\left(\frac{-(x-\gamma)}{\lambda}\right)\right]^{-2}\exp\left(\frac{-(x-\gamma)}{\lambda}\right) \\ &= \frac{1}{4\lambda}\operatorname{sech}^2\left(\frac{x-\gamma}{2\lambda}\right) \end{aligned} \tag{3.4.1}$$

$$\begin{aligned} F(x) &= \left[1+\exp\left(\frac{-(x-\gamma)}{\lambda}\right)\right]^{-1} \\ &= \frac{1}{2}\left[1+\tanh\left(\frac{1}{2}\left(\frac{x-\gamma}{\lambda}\right)\right)\right] \end{aligned} \tag{3.4.2}$$

respectively.

Application of the modified inverse stereographic projection for this model defined by the one-to-one mapping $x=\tan\left(\frac{\theta}{2}\right)$, $-\pi<\theta<\pi$, leads to Stereographic Logistic Distribution on the unit circle, so it's a circular model.

3.4.1 Probability Density Function (pdf) of Stereographic Logistic Distribution

A random variable X_S on the unit circle is said to have SLGD with location parameter μ and scale parameter $\sigma>0$ denoted by SLGD (μ,σ), if the probability density function is given by

$$g(\theta)=\frac{1}{2\sigma}\sec^2\left(\frac{\theta}{2}\right)\left[1+\exp\left(\frac{-\left(\tan\left(\frac{\theta}{2}\right)-\mu\right)}{\sigma}\right)\right]^{-2}\exp\left(-\left(\frac{\tan\left(\frac{\theta}{2}\right)-\mu}{\sigma}\right)\right),$$

where $\sigma=\dfrac{\lambda}{v}>0, \mu=\dfrac{\gamma}{v}$, and $-\pi<\theta<\pi$ (3.4.3)

Here, $g(\theta)$ satisfies the conditions 1.4.1 to 1.4.3. Hence, the proposed new model SLG (μ, σ) is a circular model called the "STEREOGRAPHIC LOGISTIC MODEL."

Theorem 3.4.1

The SLGD is symmetric about $\mu=0$ and is unimodal if $\sigma < 0.5$ and bimodal if $\sigma > 0.5$.

Proof: The probability density function of the SLGD is

$$g(\theta)=\frac{1}{2\sigma}\sec^2\left(\frac{\theta}{2}\right)\left[1+\exp\left(\frac{-\left(\tan\left(\frac{\theta}{2}\right)\right)}{\sigma}\right)\right]^{-2}\exp\left(-\left(\frac{\tan\left(\frac{\theta}{2}\right)}{\sigma}\right)\right),$$

where $\sigma > 0,\ -\pi < \theta < \pi$

equivalently

$$g(\theta)=\frac{1}{8\sigma}\sec^2\left(\frac{\theta}{2}\right)\operatorname{sech}^2\left[\frac{\tan\left(\frac{\theta}{2}\right)}{2\sigma}\right]$$

Differentiating $g(\theta)$ with respect to θ,

$$g'(\theta)=\frac{1}{8\sigma}\left[\sec^2\left(\frac{\theta}{2}\right)\tan\left(\frac{\theta}{2}\right)\operatorname{sech}^2\left(\frac{\tan\left(\frac{\theta}{2}\right)}{2\sigma}\right)\right.$$

$$\left.-\frac{1}{2\sigma}\sec^4\left(\frac{\theta}{2}\right)\operatorname{sech}^2\left(\frac{\tan\left(\frac{\theta}{2}\right)}{2\sigma}\right)\tanh\left(\frac{\tan\left(\frac{\theta}{2}\right)}{2\sigma}\right)\right]$$

$$=\frac{1}{8\sigma}\sec^2\left(\frac{\theta}{2}\right)\operatorname{sech}^2\left(\frac{\tan\left(\frac{\theta}{2}\right)}{2\sigma}\right)\left[\tan\left(\frac{\theta}{2}\right)-\frac{1}{2\sigma}\sec^2\left(\frac{\theta}{2}\right)\tanh\left(\frac{\tan\left(\frac{\theta}{2}\right)}{2\sigma}\right)\right]$$

$g'(\theta)=0 \Rightarrow \theta=0, 2\pi, 4\pi,...,$ are the stationary points.

$\theta=0$ is the only stationary point, which lies in the domain of $g(\theta)$.

At $\theta=0$

$$g''(0)=\frac{1}{16\sigma}-\frac{1}{64\sigma^3}=\frac{4\sigma^2-1}{64\sigma^3}$$

$g(\theta)$ has maximum value at $\theta=0$ if and only if $g''(0)<0 \Leftrightarrow \sigma<0.5$.
$g(\theta)$ has minimum value at $\theta=0$ if and only if $g''(0)>0 \Leftrightarrow \sigma>0.5$.

Hence the SLGD is unimodal if $\sigma<0.5$ and bimodal if $\sigma>0.5$ (Figures 3.1 and 3.2).

3.4.2 Cumulative Distribution Function (cdf) of Stereographic Logistic Distribution

The cdf of SLGD is given by (Figure 3.3).

$$G(\theta)=\left[1+\exp\left(\frac{-\left(\tan\left(\frac{\theta}{2}\right)-\mu\right)}{\sigma}\right)\right]^{-1}, \text{where } \sigma>0,\ -\pi<\theta<\pi \quad (3.4.4)$$

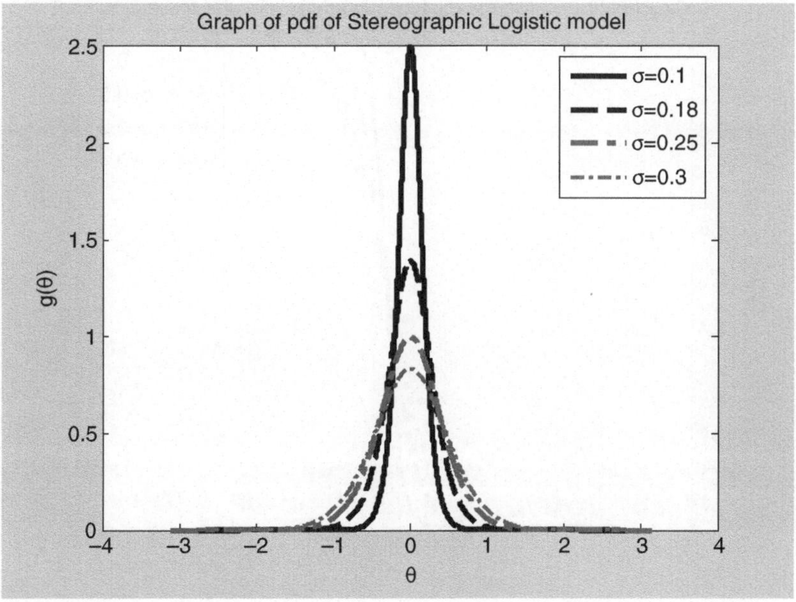

FIGURE 3.1
Pdf of stereographic logistic model for $\mu=0$ (unimodal).

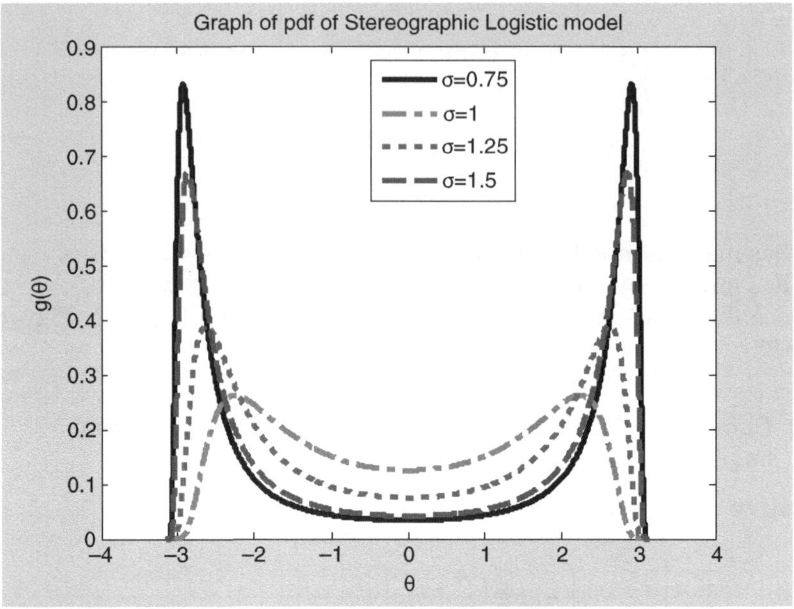

FIGURE 3.2
Graph of pdf of stereographic logistic model for $\mu = 0$ (bimodal).

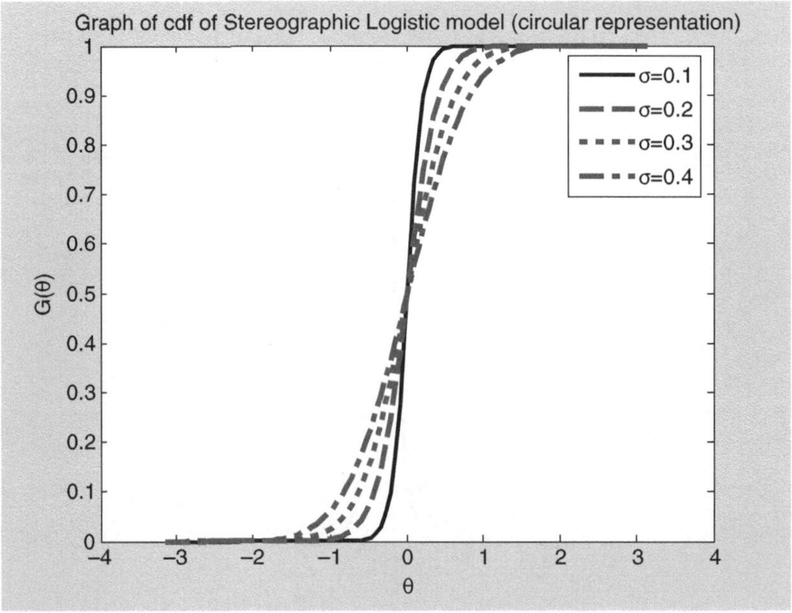

FIGURE 3.3
Graph of the cdf of stereographic logistic model for $\mu = 0$.

Role of parameter σ

$$\frac{g(\mu)}{g(\mu\pm\pi)}=\tan^2\left(\frac{\mu}{2}\right)\left[\frac{1+\exp\left(\frac{-\tan\left(\frac{\mu}{2}\right)+\mu}{\sigma}\right)}{1+\exp\left(\frac{\cot\left(\frac{\mu}{2}\right)+\mu}{\sigma}\right)}\right]^2\frac{\exp\left(\frac{-\tan\frac{\mu}{2}}{\alpha}\right)}{\exp\left(\frac{\cot\frac{\mu}{2}}{\sigma}\right)} \tag{3.4.5}$$

Hence, the larger the value of $1/\sigma$, the larger will be the ratio of $g(\mu)$ to $g(\mu\pm\pi)$ indicating higher concentration toward the population mean direction μ. Thus, $1/\sigma$ is a parameter that measures the concentration toward the mean direction.

3.4.3 Characteristic Function and Trigonometric Moments

$$\Phi_{X_S}(p)=\int_{-\pi}^{\pi}e^{ip\theta}g(\theta)\,d\theta$$

$$=\int_{-\infty}^{\infty}e^{ip\left(2\tan^{-1}\left(\frac{x}{v}\right)\right)}\frac{1}{\sigma}\left[1+e^{-\left(\frac{x}{\sigma}\right)}\right]^{-2}e^{-\left(\frac{x}{\sigma}\right)}dx$$

$$=2\frac{1}{\sigma}\int_{0}^{\infty}\cos p\left(2\tan^{-1}\left(\frac{x}{v}\right)\right)\left[1+e^{-\left(\frac{x}{\sigma}\right)}\right]^{-2}e^{-\left(\frac{x}{\sigma}\right)}dx\quad(\text{since } f(x) \text{ is even})$$

It is assumed here that $\mu=0$ in (3.4.3). The trigonometric moments of the distribution are given by $\{\varphi_p : p=\pm1,\pm2,\pm3,\ldots\}$, where $\varphi_p=\alpha_p+\beta_p$, with $\alpha_p=E(\cos p\theta)$ and $\beta_p=E(\sin p\theta)$ being the pth order cosine and sine moments of the random angle θ, respectively. Because the SLGD is symmetric about $\mu=0$, it follows that the sine moments are zero. Thus, $\varphi_p=\alpha_p$.

Theorem 3.4.2

Under the pdf of the SLGD with $\mu = 0$, the first four $\alpha_p = E(\cos p\theta)$, p=1,2,3,4, are given as follows:

$$\alpha_1 = 1 - \frac{2}{\sigma\sqrt{\pi}} \sum_{n=1}^{\infty} (-1)^{n-1} n G_{13}^{31}\left(\frac{n^2}{4\sigma^2} \,\middle|\, \begin{matrix} -\frac{1}{2} \\ -\frac{1}{2}, 0, \frac{1}{2} \end{matrix} \right),$$

$$\alpha_2 = 1 + \frac{8}{\sigma\sqrt{\pi}} \sum_{n=1}^{\infty} (-1)^{n-1} n \left\{ G_{13}^{31}\left(\frac{n^2}{4\sigma^2} \,\middle|\, \begin{matrix} -\frac{3}{2} \\ -\frac{1}{2}, 0, \frac{1}{2} \end{matrix} \right) - G_{13}^{31}\left(\frac{n^2}{4\sigma^2} \,\middle|\, \begin{matrix} -\frac{1}{2} \\ -\frac{1}{2}, 0, \frac{1}{2} \end{matrix} \right) \right\},$$

$$\alpha_3 = 1 + \frac{2}{\sigma\sqrt{\pi}} \sum_{n=1}^{\infty} (-1)^{n-1} n \left\{ \begin{matrix} 8G_{13}^{31}\left(\frac{n^2}{4\sigma^2} \,\middle|\, \begin{matrix} -\frac{5}{2} \\ -\frac{1}{2}, 0, \frac{1}{2} \end{matrix} \right) - 24G_{13}^{31}\left(\frac{n^2}{4\sigma^2} \,\middle|\, \begin{matrix} -\frac{3}{2} \\ -\frac{1}{2}, 0, \frac{1}{2} \end{matrix} \right) \\ +9G_{13}^{31}\left(\frac{n^2}{4\sigma^2} \,\middle|\, \begin{matrix} -\frac{1}{2} \\ -\frac{1}{2}, 0, \frac{1}{2} \end{matrix} \right) \end{matrix} \right\},$$

$$\alpha_4 = 1 + \frac{32}{\sigma\sqrt{\pi}} \sum_{n=1}^{\infty} (-1)^{n-1} n \left\{ \begin{matrix} \frac{2}{3}G_{13}^{31}\left(\frac{n^2}{4\sigma^2} \,\middle|\, \begin{matrix} -\frac{7}{2} \\ -\frac{1}{2}, 0, \frac{1}{2} \end{matrix} \right) + 5G_{13}^{31}\left(\frac{n^2}{4\sigma^2} \,\middle|\, \begin{matrix} -\frac{3}{2} \\ -\frac{1}{2}, 0, \frac{1}{2} \end{matrix} \right) \\ -4G_{13}^{31}\left(\frac{n^2}{4\sigma^2} \,\middle|\, \begin{matrix} -\frac{5}{2} \\ -\frac{1}{2}, 0, \frac{1}{2} \end{matrix} \right) - G_{13}^{31}\left(\frac{n^2}{4\sigma^2} \,\middle|\, \begin{matrix} -\frac{1}{2} \\ -\frac{1}{2}, 0, \frac{1}{2} \end{matrix} \right) \end{matrix} \right\},$$

where

$$\int_0^\infty x^{2v-1}\left(u^2+x^2\right)^{Q-1} e^{-\mu x}dx=\frac{u^{2v+2Q-2}}{2\sqrt{\pi}\Gamma(1-Q)}G_{13}^{31}\left(\frac{\mu^2u^2}{4}\middle|\begin{matrix}1-v\\ 1-Q-v,0,\frac{1}{2}\end{matrix}\right) \quad (3.4.6)$$

for $|\arg u\pi|<\frac{\pi}{2}$, $\mathrm{Re}\,\mu>0$ and $\mathrm{Re}\,v>0$ and $G_{13}^{31}\left(\frac{\mu^2u^2}{4}\middle|\begin{matrix}1-v\\ 1-Q-v,0,\frac{1}{2}\end{matrix}\right)$ is called as Meijer's G-function (Gradshteyn and Ryzhik, 2007, formula no. 3.389.2).

Like any other symmetric circular density, $\beta_p = E(\sin p\theta)$ are 0 as the density of the SLGD is symmetric about $\mu = 0$ (Figure 3.4).

3.4.4 Population Characteristics of SLGD

The population characteristics for the SLGD for some arbitrary values of the parameter σ are computed and tabulated (Dattatreya Rao et al. 2011). Similar tables can as well be generated for other values of the parameters. It may be noted that the first two trigonometric moments are sufficient for evaluating the population characteristics either analytically or numerically.

From the population characteristics, it can be observed that with increasing value of scale parameter σ, the circular variance gradually increases and the distribution is symmetric and starts shifting from leptokurtic to platykurtic.

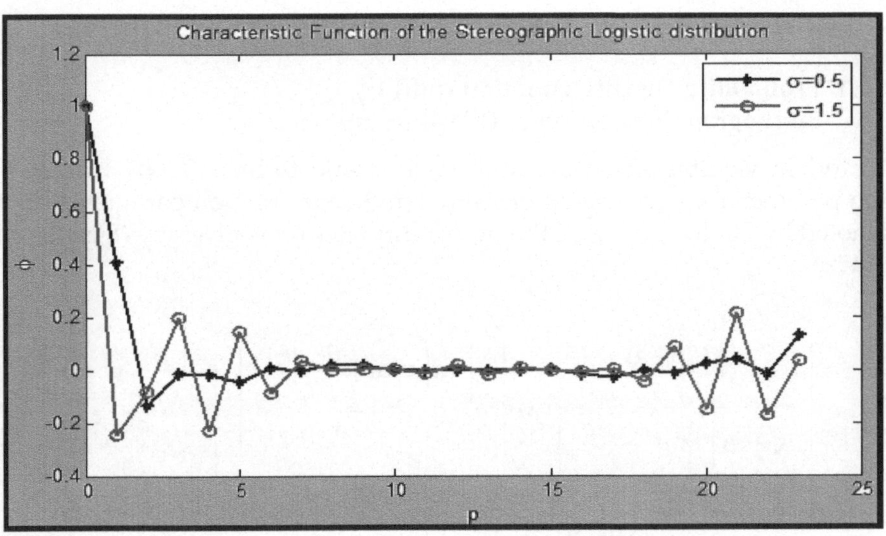

FIGURE 3.4
Graph of the characteristic function of the stereographic logistic distribution.

3.5 Stereographic Lognormal Distribution (SLND)

This section is devoted to generate a new circular model called the stereographic lognormal distribution (Girija et al. 2013) on the lines of Minh and Farnum (2003). The characteristic function for the said model is derived to evaluate the trigonometric moments, required to compute the population characteristics.

A random variable X on the real line is said to have the lognormal distribution with location parameter μ and scale parameter $\sigma > 0$, if the probability density function and the cumulative distribution function of X, respectively, are given by

$$f(x) = \begin{cases} \dfrac{1}{\sigma\sqrt{2\pi}\,(x-\mu)} \exp\left(-\left(\dfrac{\ln(x-\mu)}{\sqrt{2}\sigma}\right)^2\right), & x > \mu,\ \sigma > 0 \\ 0 & \text{if } x \le \mu \end{cases} \tag{3.5.1}$$

$$F(x) = \frac{1}{2}\left[1 + erf\left(\frac{ln(x-\mu)}{\sqrt{2}\sigma}\right)\right], \sigma > 0 \tag{3.5.2}$$

Application of the inverse stereographic projection for this model, invoking the Theorem 3.2.1 leading to the SLND on the unit circle, is a circular model.

3.5.1 Probability Density Function (pdf) of Stereographic Lognormal Distribution

A random variable X_s on the unit circle is said to have SLND with location parameter μ, scale parameter $\sigma > 0$ and concentration parameter $v > 0$, denoted by SLND (μ, σ, v), if the probability density is given by (Figures 3.5 and 3.6).

$$g(\theta) = \begin{cases} \dfrac{\sec^2\left(\dfrac{\theta-\mu}{2}\right)}{2\sqrt{2\pi}\sigma\left(\tan\left(\dfrac{\theta-\mu}{2}\right)\right)} \exp\left(-\left(\dfrac{\ln\left(v\tan\left(\dfrac{\theta-\mu}{2}\right)\right)}{\sqrt{2}\sigma}\right)^2\right) \\ 0 \qquad \text{if } \theta \le \mu \end{cases}$$

$$, v > 0,\ \sigma > 0, \mu < \theta < \pi + \mu \tag{3.5.3}$$

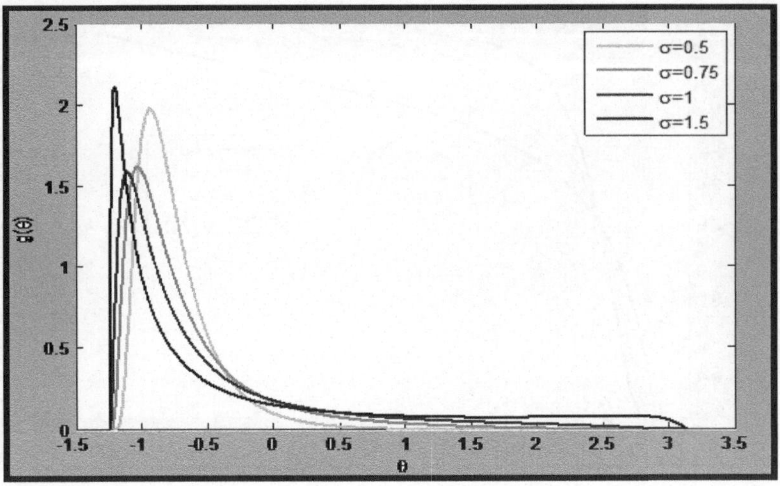

FIGURE 3.5
Pdf of the stereographic lognormal model for $v = 3.5$ and $\mu = 2.5$ (linear representation).

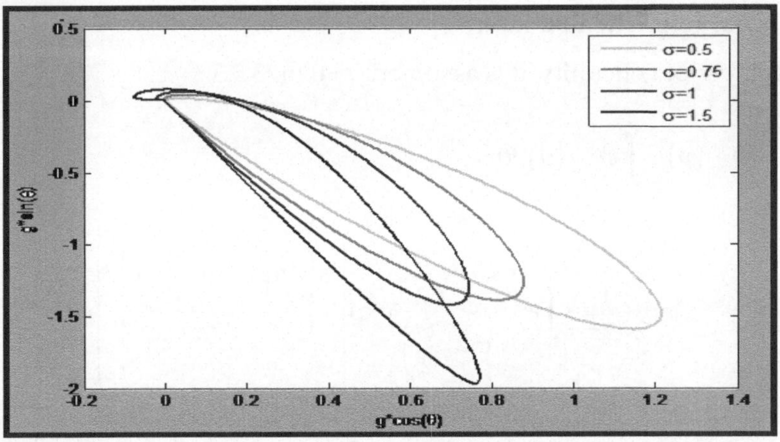

FIGURE 3.6
Pdf of the stereographic lognormal model for $v = 3.5$ and $\mu = 2.5$ (circular representation).

3.5.2 Cumulative Distribution Function (cdf) of Stereographic Lognormal Distribution

The cdf of the SLND is (Figure 3.7)

$$G(\theta) = \frac{1}{2}\left[1 + erf\left(\frac{\ln\left(v \tan\left(\frac{\theta - \mu}{2}\right)\right)}{\sqrt{2\sigma}}\right)\right], \sigma > 0, v > 0 \tag{3.5.4}$$

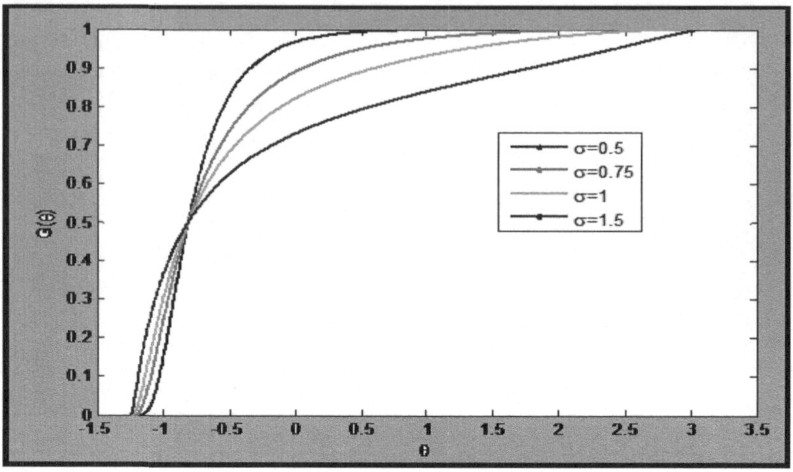

FIGURE 3.7
Cdf of the stereographic lognormal model for $v = 3.5$ and $\mu = 2.5$.

3.5.3 Characteristic Function

Without loss of generality, it is assumed $\mu=0$ in (3.5.3)

$$\Phi_{X_S}(p)=\int_{-\pi}^{\pi} e^{ip\theta} g(\theta)\,d\theta$$

$$=\frac{1}{2\sqrt{2\pi}\sigma}\int_{0}^{\pi} e^{ip\theta}\frac{\sec^2\left(\frac{\theta}{2}\right)}{\tan\left(\frac{\theta}{2}\right)}\exp\left(-\left(\frac{\ln\left(v\tan\left(\frac{\theta}{2}\right)\right)}{\sqrt{2}\sigma}\right)^2\right)d\theta$$

$$=\frac{1}{\sigma\sqrt{2\pi}}\int_{0}^{\infty}\frac{1}{x}e^{ip\left(2\tan^{-1}\left(\frac{x}{v}\right)\right)}e^{-\left(\frac{\log(x)}{\sqrt{2}\sigma}\right)^2}dx\text{, taking } x=v\tan\left(\frac{\theta}{2}\right)$$

$$=\int_{0}^{\infty} e^{ip\left(2\tan^{-1}\left(\frac{x}{v}\right)\right)} f(x)\,dx$$

$$=\varphi_{2\tan^{-1}\left(\frac{x}{v}\right)}(p)$$

As the integral cannot be obtained analytically, MATLAB techniques are applied for the evaluation of the values of the characteristic function. The graphs for real and imaginary parts of the characteristic function of the said new model are plotted here using the Gauss–Hermite integration method (Figure 3.8).

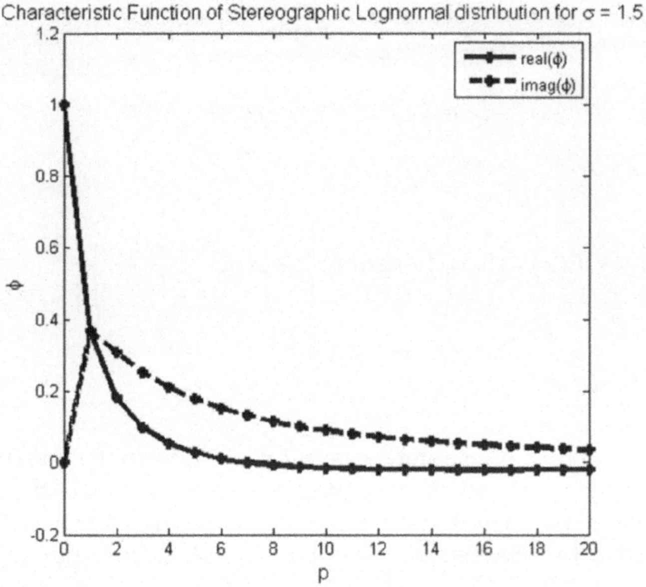

FIGURE 3.8
The characteristic function of the stereographic lognormal distribution for $\sigma = 1.5$.

3.5.4 Population Characteristics of SLND

The characteristics for the stereographic lognormal model are obtained based on Mardia and Jupp (2000) and are presented here. From the population characteristics, it can be observed that with increasing value of scale parameter σ, keeping other parameters $v = 5$, the circular variance gradually increases, the distribution is shifted from positively skewed to negatively skewed and from platykurtic to leptokurtic.

3.6 Stereographic Double-Weibull Distribution (SDWD)

New circular models are constructed by applying inverse stereographic projections on the double-Weibull (Phani et al. 2015) and the double exponential distributions, as they have got many live applications and have derived density, distribution, and characteristic functions of new circular models called the stereographic double-Weibull and the stereographic double exponential distributions, also evaluated trigonometric moments of new circular models.

A random variable X on the real line is said to have the double-Weibull distribution (Balakrishna and Kocherlakota 1985) with parameters $\lambda > 0$, $c > 0$ and μ if the probability density and the cumulative distribution functions of X are given by

$$f(x)=\frac{c\,|\,x-\alpha\,|^{c-1}}{2\lambda}\exp\left(\frac{-\,|\,x-\alpha\,|^{c}}{\lambda}\right), \tag{3.6.1}$$

$$\lambda,c>0,-\infty<\alpha<\infty,-\infty<x<\infty$$

and

$$F(x)=(0.5)\left[1+\operatorname{sgn}(x)\left\{1-\exp\left(\frac{-|x-\alpha|^{c}}{\lambda}\right)\right\}\right], \tag{3.6.2}$$

$$\lambda,c>0,\quad -\infty<x<\infty$$

respectively.

Application of the inverse stereographic projection on the lines of Theorem 3.2.1 defined by the one to one mapping $x=v\tan\left(\frac{\theta}{2}\right)$, $v>0$, $-\pi<\theta<\pi$ leads to a three-parameter symmetric circular model on the unit circle. This distribution is called the stereographic double-Weibull distribution and denoted by **SDWD** (σ,μ,c).

3.6.1 Probability Density Function (pdf) of Stereographic Double-Weibull Distribution

A random variable X_S on the unit circle is said to have the SDWD with scale parameters $\sigma>0$, shape parameter $c>0$, and location parameter μ denoted by SDW $(\sigma,\ \mu,c)$, if the probability density is given by (Figures 3.9 and 3.10)

$$g(\theta)=\frac{c}{4\sigma}\sec^{2}\left(\frac{\theta}{2}\right)\left|\tan\left(\frac{\theta}{2}\right)-\mu\right|^{c-1}\exp\left(-\frac{1}{\sigma}\left|\tan\left(\frac{\theta}{2}\right)-\mu\right|^{c}\right),$$

$$\sigma=\frac{\lambda}{v^{c}}>0,\mu=\frac{\alpha}{v},c>0,-\pi<\theta<\pi \tag{3.6.3}$$

3.6.2 Cumulative Distribution Function (cdf) of Stereographic Double-Weibull Distribution

The cdf of the SDWD distribution is (Figure 3.11)

$$G(\theta)=(0.5)\left[1+\operatorname{sgn}\left(\tan\left(\frac{\theta}{2}\right)-\mu\right)\left\{1-\exp\left(-\frac{1}{\sigma}\left|\tan\left(\frac{\theta}{2}\right)-\mu\right|^{c}\right)\right\}\right],$$

$$c,\sigma>0,-\pi<\theta<\pi \tag{3.6.4}$$

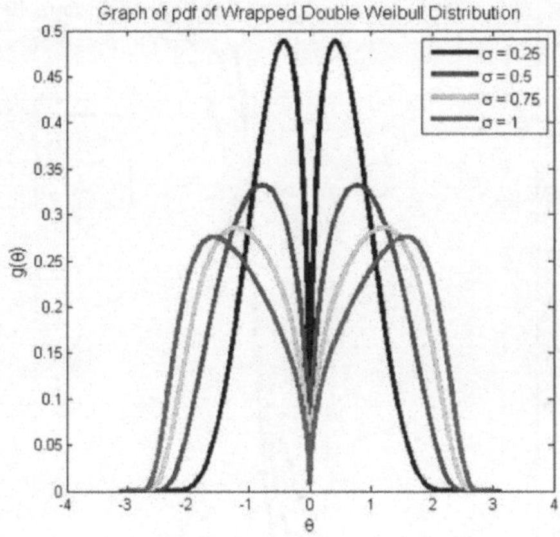

FIGURE 3.9
Pdf of the stereographic double-Weibull distribution for $c = 1.5$ (linear representation).

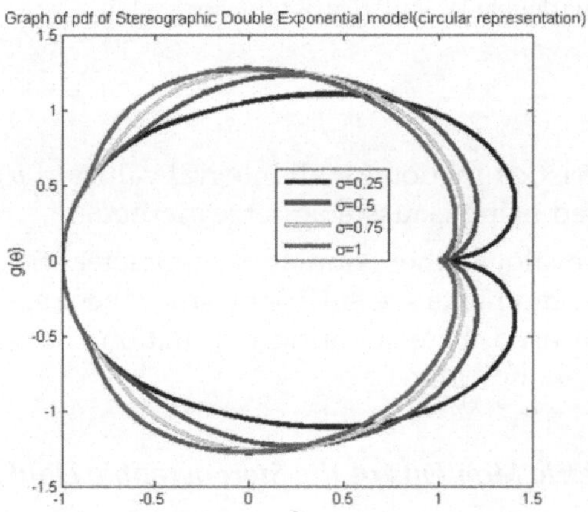

FIGURE 3.10
Pdf of the stereographic double-Weibull distribution for $c = 1.5$ (circular representation).

3.6.3 Characteristic Function and Trigonometric Moments of SDWD

$$\Phi_{X_S}(p) = \int_{-\pi}^{\pi} e^{ip\theta} g(\theta) d\theta$$

$$= \int_{-\pi}^{\pi} e^{ip\theta} \frac{c}{4\sigma} \sec^2\left(\frac{\theta}{2}\right) \left|\tan\left(\frac{\theta}{2}\right)\right|^{c-1} \exp\left(-\frac{1}{\sigma}\left|\tan\left(\frac{\theta}{2}\right)\right|^{c}\right) d\theta$$

$$= \frac{c}{2\sigma} \int_{0}^{\pi} \cos p\theta \sec^2\left(\frac{\theta}{2}\right) \left(\tan\left(\frac{\theta}{2}\right)\right)^{c-1} \exp\left(-\frac{1}{\sigma}\left(\tan\left(\frac{\theta}{2}\right)\right)^{c}\right) d\theta$$

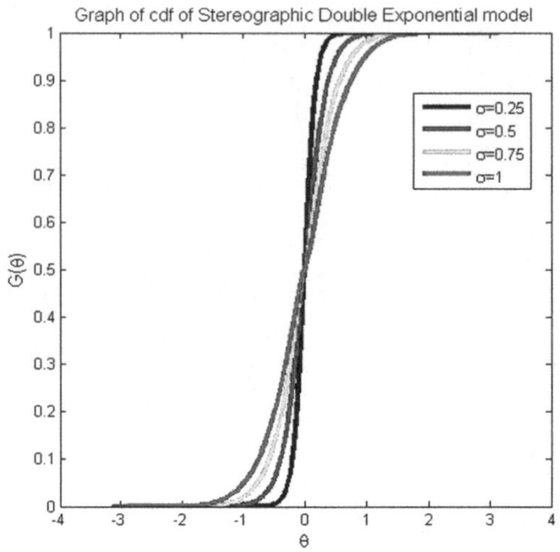

FIGURE 3.11
Cdf of the stereographic double-Weibull distribution for $c = 1.5$.

NOTE:

1. The characteristic function for all integral values of p can be numerically evaluated using Gauss–Laguerre method.
2. In order to evaluate the population characteristics, the first two trigonometric moments are sufficient, since these are in closed form and could be used for evaluating population characteristics either numerically/analytically.

3.6.3.1 Trigonometric Moments of the Stereographic Double-Weibull Model

The trigonometric moments of the distribution are given by $\{\varphi_p : p = \pm 1, \pm 2, \pm 3, \ldots\}$, where $\varphi_p = \alpha_p + \beta_p$, with $\alpha_p = E(\cos p\theta)$ and $\beta_p = E(\sin p\theta)$ being the pth order cosine and sine moments of the random angle θ, respectively. Because the SDWD is symmetric about $\mu = 0$, it follows that the sine moments are zero. Thus, $\varphi_p = \alpha_p$.

Theorem 3.6.1

Under the pdf of the SDWD with $\mu = 0$, the first four $\alpha_p = E(\cos p\theta)$, $p = 1, 2, 3, 4$, are given as follows:

$$\alpha_1 = 1 - \frac{2}{\sigma} \sum_{n=0}^{\infty} (-1)^n (\sigma)^{\left(\frac{2n+c+2}{c}\right)} \Gamma\left(\frac{2n+c+2}{c}\right),$$

$$\alpha_2 = 1 + \frac{4}{\sigma}\sum_{n=0}^{\infty}(-1)^n (n+1)(\sigma)^{\left(\frac{2n+c+4}{c}\right)}\Gamma\left(\frac{2n+c+4}{c}\right)$$

$$-\frac{4}{\sigma}\sum_{n=0}^{\infty}(-1)^n (\sigma)^{\left(\frac{2n+c+2}{c}\right)}\Gamma\left(\frac{2n+c+2}{c}\right),$$

$$\alpha_3 = 1 - \frac{16}{\sigma}\sum_{n=0}^{\infty}(-1)^n (n+1)(n+2)(\sigma)^{\left(\frac{2n+c+6}{c}\right)}\Gamma\left(\frac{2n+c+6}{c}\right)$$

$$+\frac{48}{\sigma}\sum_{n=0}^{\infty}(-1)^n (\sigma)^{\left(\frac{2n+c+4}{c}\right)}\Gamma\left(\frac{2n+c+4}{c}\right)$$

$$-\frac{18}{\sigma}\sum_{n=0}^{\infty}(-1)^n (\sigma)^{\left(\frac{2n+c+2}{c}\right)}\Gamma\left(\frac{2n+c+2}{c}\right),$$

$$\alpha_4 = 1 + \frac{64}{3\sigma}\sum_{n=0}^{\infty}(-1)^n (n+1)(n+2)(n+3)(\sigma)^{\left(\frac{2n+c+8}{c}\right)}\Gamma\left(\frac{2n+c+8}{c}\right)$$

$$+\frac{160}{\sigma}\sum_{n=0}^{\infty}(-1)^n (n+1)(\sigma)^{\left(\frac{2n+c+1}{c}\right)}\Gamma\left(\frac{2n+c+1}{c}\right)$$

$$-\frac{128}{\sigma}\sum_{n=0}^{\infty}(-1)^n (n+1)(n+2)(\sigma)^{\left(\frac{2n+c+1}{c}\right)}\Gamma\left(\frac{2n+c+1}{c}\right)$$

$$-\frac{32}{\sigma}\sum_{n=0}^{\infty}(-1)^n (\sigma)^{\left(\frac{2n+c+2}{c}\right)}\Gamma\left(\frac{2n+c+2}{c}\right)$$

where

$$\int_0^{\infty} x^{\nu-1} e^{\left(-\mu x^p\right)} dx = \frac{1}{p}\mu^{-\frac{\nu}{p}}\Gamma\left(\frac{\nu}{p}\right) \text{ for Re}\,\mu > 0, \text{Re}\,\nu > 0 \text{ and } p > 0 \qquad (3.6.5)$$

(Gradshteyn and Ryzhik, 2007, formula no. 3.478.1).

3.6.4 Population Characteristics

Using mathematical expressions in Mardia and Jupp (2000) and the first two trigonometric moments, the characteristics of the SDWD are computed. From the population characteristics, it can be observed that with increasing value of scale parameter σ, keeping shape parameter c = 1.5, the circular variance gradually increases, the distribution is symmetric and it shifts from leptokurtic to platykurtic.

3.7 Other Stereographic Circular Models

Some of the other stereographic circular models constructed by Phani et al. (2012a, 2012b) are presented here.

3.7.1 Stereographic Extreme-Value Distribution (SEVD) (Phani et al. 2012a)

a. Probability Density Function

$$g(\theta)=\frac{1}{2\sigma}\sec^2\left(\frac{\theta}{2}\right)\exp\left(-\frac{\left(\tan\left(\frac{\theta}{2}\right)-\mu\right)}{\sigma}\right)\exp\left\{-\exp\left(-\frac{\left(\tan\left(\frac{\theta}{2}\right)-\mu\right)}{\sigma}\right)\right\}, \tag{3.7.1}$$

where, $\mu=\frac{\gamma}{v}$, $\sigma=\frac{\lambda}{v}>0$

b. Cumulative Distribution Function

$$G(\theta)=\exp\left\{-\exp\left(-\frac{\left(\tan\left(\frac{\theta}{2}\right)-\mu\right)}{\sigma}\right)\right\}, \text{where } \sigma>0, -\pi<\theta<\pi. \tag{3.7.2}$$

c. Characteristic Function

$$\Phi_{X_S}(p)=\int_{-\pi}^{\pi}e^{ip\theta}g(\theta)d\theta$$

$$=\frac{1}{2\sigma}\int_{-\pi}^{\pi}e^{ip\theta}\sec^2\left(\frac{\theta}{2}\right)e^{-\frac{\tan\left(\frac{\theta}{2}\right)}{\sigma}}e^{-e^{-\left(\frac{\tan\left(\frac{\theta}{2}\right)}{\sigma}\right)}}d\theta$$

$$=\frac{1}{\sigma}\int_{-\infty}^{\infty}e^{ip\left(2\tan^{-1}\left(\frac{x}{v}\right)\right)}e^{-\left(\frac{x}{\sigma}\right)}e^{-e^{-\left(\frac{x}{\sigma}\right)}}dx, \text{ taking } x=\tan\left(\frac{\theta}{2}\right)$$

The characteristics for the **stereographic extreme-value model** are obtained based on Mardia and Jupp (2000).

3.7.2 Stereographic Reflected Gamma Distribution (SRGD) (Phani et al. 2012b)

a. Probability Density Function

$$g(\theta)=\frac{1}{4\sigma^c\Gamma(c)}\sec^2\left(\frac{\theta}{2}\right)\left|\tan\left(\frac{\theta}{2}\right)-\mu\right|^{c-1}\exp\left(-\frac{1}{\sigma}\left|\tan\left(\frac{\theta}{2}\right)-\mu\right|\right),$$

where

$$c,\sigma=\frac{\lambda}{v}>0,\ \mu=\frac{\alpha}{v},-\pi<\theta<\pi \tag{3.7.3}$$

b. Cumulative Distribution Function

$$G(\theta)=\begin{cases}\frac{1}{2}\left(1-\frac{\Gamma_{x_1}(c)}{\Gamma(c)}\right), \text{ where } x_1=-\frac{1}{\sigma}\tan\left(\frac{\theta}{2}\right) \text{ for } \theta\in(-\pi,0)\\ \frac{1}{2}\left(1+\frac{\Gamma_{x_2}(c)}{\Gamma(c)}\right), \text{ where } x_2=\frac{1}{\sigma}\tan\left(\frac{\theta}{2}\right) \text{ for } \theta\in[0,\pi)\end{cases}$$

$$c,\sigma>0,\ -\pi<\theta<\pi \tag{3.7.4}$$

c. Characteristic Function of the Stereographic Reflected Gamma Distribution

$$\varphi_{X_S}(p)=\int_{-\pi}^{\pi}e^{ip\theta}g(\theta)d\theta$$

$$=\frac{1}{4\sigma^c\Gamma(c)}\int_{-\pi}^{\pi}e^{ip\theta}\sec^2\left(\frac{\theta}{2}\right)\left|\tan\left(\frac{\theta}{2}\right)\right|^{c-1}e^{-\frac{1}{\sigma}\left|\tan\left(\frac{\theta}{2}\right)\right|}d\theta$$

$$=\frac{1}{2\sigma^c\Gamma(c)}\int_{0}^{\pi}e^{ip\theta}\sec^2\left(\frac{\theta}{2}\right)\tan\left(\frac{\theta}{2}\right)^{c-1}e^{-\frac{1}{\sigma}\tan\left(\frac{\theta}{2}\right)}d\theta$$

$$=\frac{1}{2\sigma^c\Gamma(c)}\int_{0}^{\pi}\cos(p\theta)\sec^2\left(\frac{\theta}{2}\right)\tan\left(\frac{\theta}{2}\right)^{c-1}e^{-\frac{1}{\sigma}\tan\left(\frac{\theta}{2}\right)}d\theta \tag{3.7.5}$$

References

Abramowitz, M. and Stegun, I. A. 1965. *Handbook of Mathematical Functions*, Dover, New York.

Balakrishnan, N. and Kocherlakota, S. 1985. On the Double Weibull Distribution: Order Statistics and Estimation, *Sankhya, B – Series*, 47(2), 161–178.

Dattatreya Rao, A.V., Girija, S.V.S. and Phani, Y. 2011. On Stereographic Logistic Model, *Proceedings of NCAMES*, AU Engineering College, Visakhapatnam, India, pp. 139–141.

Dattatreya Rao, A.V., Girija, S.V.S. and Phani, Y. 2013. Arc Tan-Exponential Type Distribution Induced by Stereographic Projection/Bilinear Transformation on Modified Wrapped Exponential Distribution, *Journal of the Applied Mathematics, Statistics and Informatics (JAMSI)*, 9(1), 69–74.

Dattatreya Rao, A.V., Girija, S.V.S. and Phani, Y. 2016. Stereographic Logistic Model: Application to Noisy Scrub Birds Data, *Chilean Journal of Statistics*, 7(2), 69–79.

Girija, S.V.S., Dattatreya Rao, A.V. and Phani, Y. 2013. On Stereographic Lognormal Distribution, *International Journal of Advances in Applied Sciences (IJAAS)*, 2(3), 125–132.

Gradshteyn, I.S. and Ryzhik, I.M. 2007. *Table of Integrals, Series and Products*, 7th ed., Academic Press, London, UK.

Jammalamadaka, S.R. and Sen Gupta, A. 2001. *Topics in Circular Statistics*, World Scientific Press, Singapore.

Lukacs, E. 1970. *Characteristic Functions*, 2nd ed., Griffin, London, UK.

Mardia, K. V. and Jupp, P. E. 2000. *Directional Statistics*, John Wiley & Sons, Chichester, UK.

Minh, D.P. and Farnum, N.R. 2003. Using Bilinear Transformation to Induce Probability Distributions, *Communications in Statistics-Theory and Methods*, 32(1), 1–9.

Phani, Y. 2013. On Stereographic Circular and Semicircular Models, PhD dissertation, Acharya Nagarjuna University, Guntur, India.

Phani, Y., Girija, S.V.S. and Dattatreya Rao, A.V. 2012a, Circular Model Induced by Inverse Stereographic Projection on Extreme-Value Distribution, *IRACST: Engineering Science and Technology: An International Journal (ESTIJ)*, 2(5), 881–888.

Phani, Y., Radhika, A.J.V., Girija, S.V.S. and Dattatreya Rao, A.V. 2012b. Modeling Ants Data Using Stereographic Reflected Gamma Distribution, *ANU Journal of Physical Sciences*, 4, 15–38.

Phani, Y., Girija, S.V.S. and Dattatreya Rao, A.V. 2015. Symmetric Circular Model Induced by Inverse Stereographic Projection on Double Weibull Distribution with Application, *International Journal of Soft Computing, Mathematics and Control (IJSCMC)*, 4(1), 69–76.

Ramabhadra Sarma, I., Dattatreya Rao, A.V. and Girija, S.V.S. 2009. On Characteristic Functions of Wrapped Half Logistic and Binormal Distributions, *International Journal of Statistics and Systems*, 4(1), 33–45.

Ramabhadra Sarma, I., Dattatreya Rao, A.V. and Girija, S.V.S. 2011. On Characteristic Functions of Wrapped Lognormal and Weibull Distributions, *Journal of Statistical Computation and Simulation*, 81(5), 579–589.

Toshihiro, A., Pewsey, A. and Kunio, S. 2013. Extending Circular Distributions Through Transformation of Argument, *Annals of the Institute of Statistical Mathematics*, 65(5), 833–858.

4

Offset Circular Models

4.1 Introduction

Most of the existing circular models that appear in the literature are unimodal. However, bimodal data (distributions) have occurred in many practical applications, such as

1. Genome expression (of RNA) data (Mason et al. 2011)
2. X-ray diffraction characterization of size and shape distributions of nanoparticles (Armstrong et al. 2004)
3. The majority-vote dynamics associated with each spin in 2D square lattice, the data of the noise parameter (Vilela et al. 2012)
4. Thirteen homing pigeon problem (Jammalamadaka and Sen Gupta 2001)

It is indicated that the angular data set of 13 homing pigeon problem (Jammalamadaka and Sen Gupta 2001, p. 165) is somewhat bimodal and established that offset Pearson-type II distribution (OP-II) is a good fit.

Hence, this chapter is devoted to discussing construction of bimodal circular models using the technique of offsetting. The process of transforming a bivariate linear random variable to its directional component is called offsetting, and the resulting distribution is a univariate circular model.

Applying the method of offsetting new bimodal circular models called offset Cauchy distribution (OCD), OP-II, and offset t-distribution (OTD) are derived in this chapter. Also included are the graphs of pdf, cdf, and characteristic functions of the above new models.

4.2 Methodology of Offsetting

As mentioned earlier, the method of transforming a bivariate linear r.v. to its directional component is called OFFSETTING, and the respective distribution of directional component is called offset distribution. This is done by accumulating probabilities over all different lengths for a given direction. Jammalamadaka and Sen Gupta (2001) established that the bivariate random vector (X, Y) can be transformed into polar coordinates (R,θ), and when it is integrated over R for a given θ yields a univariate circular model. If $f(x, y)$ denotes the joint density distribution on the plane, then the resulting circular offset distribution, say $g(\theta)$, is given by

$$g(\theta)=\int_0^\infty f(r\cos\theta, r\sin\theta)\,r\,dr \tag{4.2.1}$$

4.3 Offset Cauchy Model

The bivariate Cauchy distribution is applied in many areas, including biological analyses, clinical trials, stochastic modeling of decreasing failure rate life components, study of labor turnover, queuing theory, and reliability (Nayak 1987, Lee and Gross 1991). In the study of biological analyses, clinical trials, and reliability, the circular distributions will yield better results. The Offset Cauchy Distribution (OCD) (Girija et al. 2013b) is derived from the bivariate Cauchy distribution (Balakrishnan and Chin-Diew 2008, p. 365).

4.3.1 Probability Density Function (pdf) of Offset Cauchy Distribution

The pdf $g(\theta)$ of the offset Cauchy model for the bivariate Cauchy distribution with positive parameters σ_1 and σ_2 is given by

$$g(\theta)=\frac{\sigma_1\sigma_2}{2\pi\left(\sigma_2^2\cos^2\theta+\sigma_1^2\sin^2\theta\right)}, 0\le\theta<2\pi \tag{4.3.1}$$

Proof: The joint density function of the bivariate Cauchy distribution with parameters σ_1 and σ_2 is

$$f(x,y)=\frac{1}{2\pi\sigma_1\sigma_2}\left(1+\frac{x^2}{\sigma_1^2}+\frac{y^2}{\sigma_2^2}\right)^{-3/2} \text{ where } x,y\in\mathbb{R}, \sigma_1,\sigma_2>0 \tag{4.3.2}$$

Using the technique explained in Section 4.2, the probability density function of the OCD is derived as under

$$g(\theta)=\int_0^\infty f(r\cos\theta, r\sin\theta)\,r\,dr$$

$$=\int_0^\infty \frac{1}{2\pi\sigma_1\sigma_2}\left(1+\frac{r^2\cos^2\theta}{\sigma_1^2}+\frac{r^2\sin^2\theta}{\sigma_2^2}\right)^{-3/2} r\,dr,\ \left(\text{put } r^2=t\right)$$

$$=\frac{\sigma_1\sigma_2}{2\pi\left(\sigma_2^2\cos^2\theta+\sigma_1^2\sin^2\theta\right)} \qquad \text{where } \sigma_1>\sigma_2>0,\ \theta\in[0,2\pi)$$

This function satisfies the properties of circular model 1.4.1 through 1.4.3. Hence, clearly g is a pdf of the circular model.

Hence, the function $g(\theta)$ represents a probability density function of a circular model called the OCD. When $\sigma_1=\sigma_2$, the density reduces to the circular uniform density.

The linear representation of the graph of the pdf of the OCD for various values of σ_1 and σ_2 can be plotted using MATLAB® (Figure 4.1).

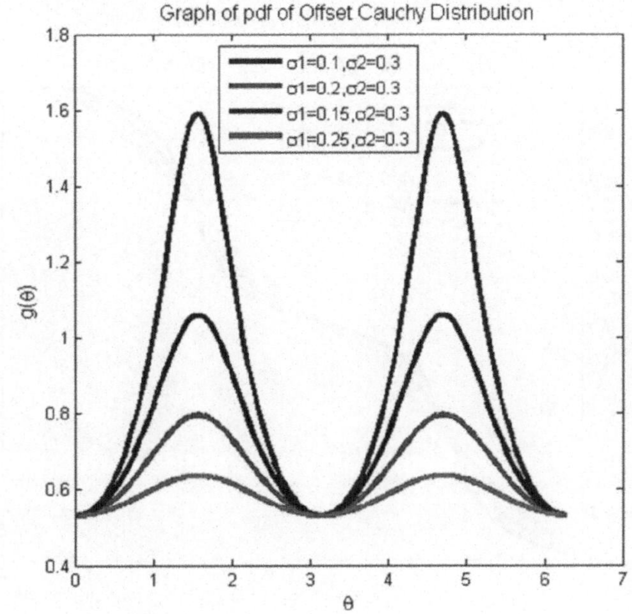

FIGURE 4.1
Graph of pdf of the offset Cauchy distribution (bimodal).

4.3.2 Cumulative Distributive Function (cdf) of Offset Cauchy Distribution

The cumulative distribution function (cdf) $G(\theta)$ is given by (Figure 4.2)

$$G(\theta)=\int_0^{\theta} g(\alpha)\,d\alpha=\int_0^{\theta}\frac{\sigma_1\sigma_2}{2\pi\left(\sigma_2^2\cos^2\alpha+\sigma_1^2\sin^2\alpha\right)}\,d\alpha$$

$$=\frac{\sigma_1\sigma_2}{2\pi}\int_0^{\tan\theta}\frac{dt}{\left(\sigma_2^2+\sigma_1^2t^2\right)}\qquad (\text{put } \tan\alpha = \text{t})$$

$$=\begin{cases}\dfrac{1}{2\pi}\tan^{-1}\left(\dfrac{\tan\theta}{\sigma_2/\sigma_1}\right) & \text{for } 0\le\theta<\dfrac{\pi}{2}\\[2ex] \dfrac{1}{2\pi}\left\{\pi+\tan^{-1}\left(\dfrac{\tan\theta}{\sigma_2/\sigma_1}\right)\right\} & \text{for } \dfrac{\pi}{2}<\theta<\dfrac{3\pi}{2}\\[2ex] \dfrac{1}{2\pi}\left\{2\pi+\tan^{-1}\left(\dfrac{\tan\theta}{\sigma_2/\sigma_1}\right)\right\} & \text{for } \dfrac{3\pi}{2}<\theta<2\pi\end{cases}\qquad (4.3.3)$$

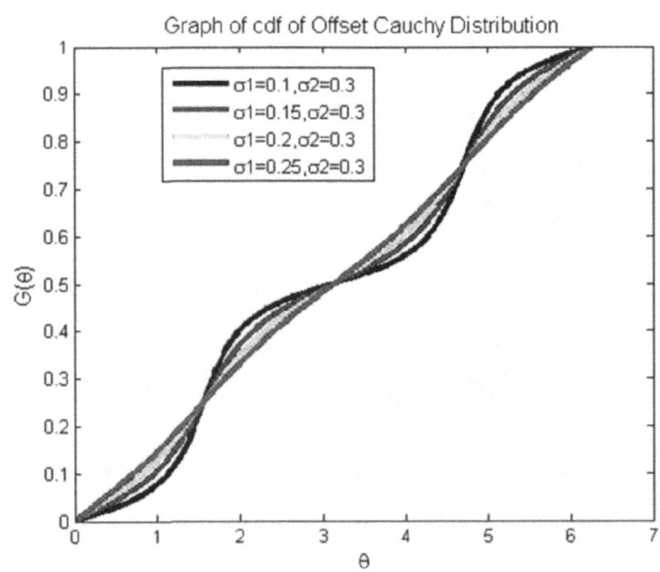

FIGURE 4.2
Graph of cdf of the offset Cauchy distribution.

A random variable X_O on the unit circle is said to have OCD with parameters $\sigma_1,\sigma_2 > 0$ denoted by OCD (σ_1,σ_2), if the probability density and cumulative distribution functions are respectively given by

$$g(\theta) = \frac{\sigma_1 \sigma_2}{2\pi\left(\sigma_2^2 \cos^2\theta + \sigma_1^2 \sin^2\theta\right)}, 0 \le \theta < 2\pi$$

$$G(\theta)=\begin{cases} \frac{1}{2\pi}\tan^{-1}\left(\frac{\tan\theta}{\sigma_2/\sigma_1}\right) & \text{for } 0 \le \theta < \frac{\pi}{2} \\ \frac{1}{2\pi}\left\{\pi + \tan^{-1}\left(\frac{\tan\theta}{\sigma_2/\sigma_1}\right)\right\} & \text{for } \frac{\pi}{2} < \theta < \frac{3\pi}{2} \\ \frac{1}{2\pi}\left\{2\pi + \tan^{-1}\left(\frac{\tan\theta}{\sigma_2/\sigma_1}\right)\right\} & \text{for } \frac{3\pi}{2} < \theta < 2\pi \end{cases}$$

4.3.3 Characteristic Function and Trigonometric Moments

The characteristic function of the OCD is derived on the lines of Ramabhadra Sarma et al. (2009, 2011). By applying the Cauchy Residue Theorem, the characteristic function of the OCD is obtained as

$$\varphi_p = \int_0^{2\pi} e^{ip\theta}\, g(\theta)\, d\theta$$

$$= \int_0^{2\pi} \frac{e^{ip\theta}\ \sigma_1\ \sigma_2}{2\pi\ \left(\sigma_2^2 \cos^2\theta + \sigma_1^2 \sin^2\theta\right)}\ d\theta$$

$$= \frac{2\ \sigma_1\ \sigma_2}{\pi\ i} \int_C \frac{z^{p+1}\ dz}{z^4\left(\sigma_2^2 - \sigma_1^2\right) + 2z^2\left(\sigma_2^2 + \sigma_1^2\right) + \left(\sigma_2^2 - \sigma_1^2\right)}$$

By applying Cauchy's Residue Theorems (Chillingworth 1973)

$$\int_C f(z)\, dz = 2\pi i \left\{\text{Sum of residues at poles within } |z| = 1\right\}$$

$$= 2\,\pi\ i\left\{\frac{2\sigma_1\sigma_2}{\pi\ i\left(\sigma_2^2 - \sigma_1^2\right)}\ \frac{a^p\left(1+(-1)^p\right)}{2\left(a^2 - b^2\right)}\right\}$$

$$\int_C f(z)dz = \frac{a^p\left(1+(-1)^p\right)}{2}$$

$$\begin{aligned}\varphi_\theta(p) &= \int_0^{2\pi} e^{ip\theta}\frac{\sigma_1\sigma_2}{2\pi\left(\sigma_2^2\cos^2\theta+\sigma_1^2\sin^2\theta\right)}d\theta,\ p\in\mathbb{Z} \\ &= \frac{a^p\left(1+(-1)^p\right)}{2} \qquad \text{where } a=\sqrt{\frac{\sigma_1-\sigma_2}{\sigma_1+\sigma_2}}\end{aligned} \tag{4.3.4}$$

The Fourier representation of the pdf of the circular model is (Jammalamadaka and Sen Gupta 2001)

$$\begin{aligned}g(\theta) &= \frac{1}{2\pi}\sum_{-\infty}^{\infty}\varphi_p e^{-ip\theta} \\ &= \frac{1}{2\pi}\left[1+2\sum_{p=1}^{\infty}(\alpha_p\cos p\theta+\beta_p\sin p\theta)\right]\ \theta\in[0,2\pi),p\in\mathbb{Z}\end{aligned} \tag{4.3.5}$$

where the convergence of the sum is in the L^2 sense.

Here, in the case of the offset Cauchy model, it is shown as

$$\begin{aligned}g(\theta) &= \frac{1}{2\pi}\left[1+2\sum_{p=1}^{\infty}a^p\cos p\theta\right] \\ &= \frac{1}{2\pi}\left[1+2\sum_{k=1}^{\infty}a^{2k}\cos 2k\theta\right] \qquad \text{where } a=\sqrt{\frac{\sigma_1-\sigma_2}{\sigma_1+\sigma_2}} \\ &= \frac{1}{\pi}\left[\sum_{k=0}^{\infty}a^{2k}\cos 2k\theta-\frac{1}{2}\right]\end{aligned} \tag{4.3.6}$$

It is observed that the OCD is symmetric and bimodal only if σ_1 and σ_2 are unequal. The value of the characteristic function at an integer p is called the pth trigonometric moment of θ. Here, first trigonometric moments α_1 and β_1 are zeros. Hence, properties of the distribution can't be evaluated by the formulae available in Mardia and Jupp (2000). The mean is zero and the variance is one (Jammalamadaka and Sen Gupta 2001, p. 43).

4.4 Offset Pearson-Type II Model

The bivariate Pearson-type II distribution arises in many statistical problems, including analysis of variance and experimental design in general and two-stage estimation procedures, but is rarely used to fit data. The Offset Pearson Type II Model (OP-II) distribution (Radhika et al. 2013a) is derived from the bivariate Pearson-type II distribution (Balakrishnan and Chin-Diew 2008, p. 371).

4.4.1 Probability Density Function (pdf) of Offset Pearson-Type II Model

The pdf $g(\theta)$ of the OP-II model for the bivariate Pearson-type II distribution with parameters $q > 1$ and ρ where $|\rho| < 1$ is given by

$$g(\theta) = \frac{\sqrt{1-\rho^2}}{2\pi\,(1-\rho\sin 2\theta)} \quad \text{where } \theta \in [0, 2\pi) \tag{4.4.1}$$

Proof: The joint density function of Bivariate Pearson Type II distribution is

$$f(x,y) = \frac{q+1}{\pi\sqrt{1-\rho^2}}\left[1 - \frac{x^2 - 2\rho xy + y^2}{1-\rho^2}\right]^q \tag{4.4.2}$$

where $q > 1$, $|\rho| < 1$ and (x, y) is in the ellipse $x^2 - 2\rho xy + y^2 = 1 - \rho^2$, which itself lies within the unit square.

The probability density function of the OP-II distribution is derived as under

$$g(\theta) = \int_0^{\infty} f(r\cos\theta, r\sin\theta)\, r\, dr$$

Substituting $x = r\cos\theta$, $y = r\sin\theta$ in $x^2 - 2\rho xy + y^2 = 1 - \rho^2$, then $r = \frac{\sqrt{1-\rho^2}}{1-\rho\sin 2\theta}$

$$g(\theta) = \frac{q+1}{\pi\sqrt{1-\rho^2}} \int_0^{r}\left(1 - \frac{(1-\rho\sin 2\theta)r^2}{1-\rho^2}\right)^q r\, dr$$

$$= \frac{\sqrt{1-\rho^2}}{2\pi\,(1-\rho\sin 2\theta)}$$

4.4.2 Cumulative Distribution Function (cdf) of Offset Pearson-Type II Model

The cdf $G(\theta)$ of the OP-II distribution is

$$G(\theta)=\begin{cases}\dfrac{1}{2\pi}\left\{\tan^{-1}\left(\dfrac{\tan\theta-\rho}{\sqrt{1-\rho^2}}\right)-\tan^{-1}\left(\dfrac{-\rho}{\sqrt{1-\rho^2}}\right)\right\}, \theta\in\left[0,\dfrac{\pi}{2}\right) \\ \dfrac{1}{2\pi}\left\{\pi+\tan^{-1}\left(\dfrac{\tan\theta-\rho}{\sqrt{1-\rho^2}}\right)-\tan^{-1}\left(\dfrac{-\rho}{\sqrt{1-\rho^2}}\right)\right\}, \theta\in\left(\dfrac{\pi}{2},\dfrac{3\pi}{2}\right) \\ \dfrac{1}{2\pi}\left\{2\pi+\tan^{-1}\left(\dfrac{\tan\theta-\rho}{\sqrt{1-\rho^2}}\right)-\tan^{-1}\left(\dfrac{-\rho}{\sqrt{1-\rho^2}}\right)\right\}, \theta\in\left(\dfrac{3\pi}{2},2\pi\right)\end{cases} \quad (4.4.3)$$

A random variable X_O on the unit circle is said to have OP-II distribution with parameter $|\rho|<1$ denoted by OP-II (ρ) if the probability density and cumulative distribution functions are respectively given by (Figures 4.3 and 4.4)

$$g(\theta)=\frac{\sqrt{1-\rho^2}}{2\pi\,(1-\rho\sin 2\theta)} \quad \text{where } \theta\in[0,2\pi)$$

and

$$G(\theta)=\begin{cases}\dfrac{1}{2\pi}\left\{\tan^{-1}\left(\dfrac{\tan\theta-\rho}{\sqrt{1-\rho^2}}\right)-\tan^{-1}\left(\dfrac{-\rho}{\sqrt{1-\rho^2}}\right)\right\}, \theta\in\left[0,\dfrac{\pi}{2}\right) \\ \dfrac{1}{2\pi}\left\{\pi+\tan^{-1}\left(\dfrac{\tan\theta-\rho}{\sqrt{1-\rho^2}}\right)-\tan^{-1}\left(\dfrac{-\rho}{\sqrt{1-\rho^2}}\right)\right\}, \theta\in\left(\dfrac{\pi}{2},\dfrac{3\pi}{2}\right) \\ \dfrac{1}{2\pi}\left\{2\pi+\tan^{-1}\left(\dfrac{\tan\theta-\rho}{\sqrt{1-\rho^2}}\right)-\tan^{-1}\left(\dfrac{-\rho}{\sqrt{1-\rho^2}}\right)\right\}, \theta\in\left(\dfrac{3\pi}{2},2\pi\right)\end{cases}.$$

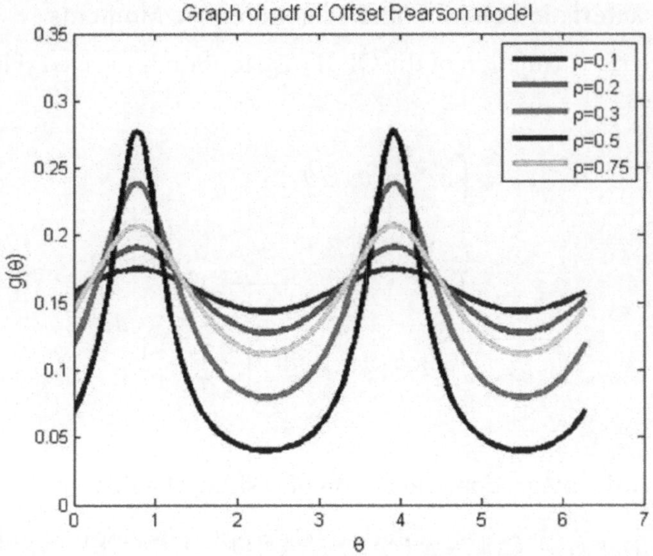

FIGURE 4.3
Graph of pdf of the offset Pearson-type II model (bimodal).

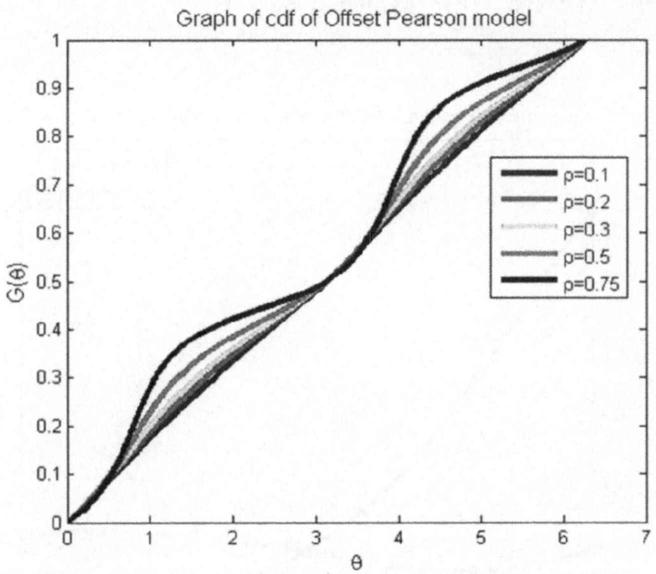

FIGURE 4.4
Graph of cdf of the offset Pearson-type II model.

4.4.3 Characteristic Function and Trigonometric Moments

The characteristic function of the OP-II distribution is derived (Figure 4.5).

$$\varphi_p = \int_0^{2\pi} e^{ip\theta}\; g(\theta)\; d\theta$$

$$= \int_0^{2\pi} \frac{e^{ip\theta}\sqrt{1-\rho^2}}{2\pi\sigma\left(1-\rho\sin 2\left(\frac{\theta-\mu}{\sigma}\right)\right)}\, d\theta$$

To evaluate this integration, the notion of residue at a finite point is applied. $f(z)=\frac{\sqrt{1-\rho^2}}{2\pi i}\left(\frac{e^{ip\mu}z^{\sigma p-1}}{1-\frac{\rho}{2i}\left(z^2-\frac{1}{z^2}\right)}\right)$ is analytic in C except at its singular points within C, so that $\int_C f(z)\,dz = 2\pi i$ (sum of residues at the singular points within C). If α^2 and β^2 are the roots of the denominator, then the poles of *f* within *C* are $\pm\beta$, where $\beta^2 = \left(\frac{\left(1-\sqrt{1-\rho^2}\right)}{\rho}i\right)$ and clearly $\left|\alpha^2\right|>1$.

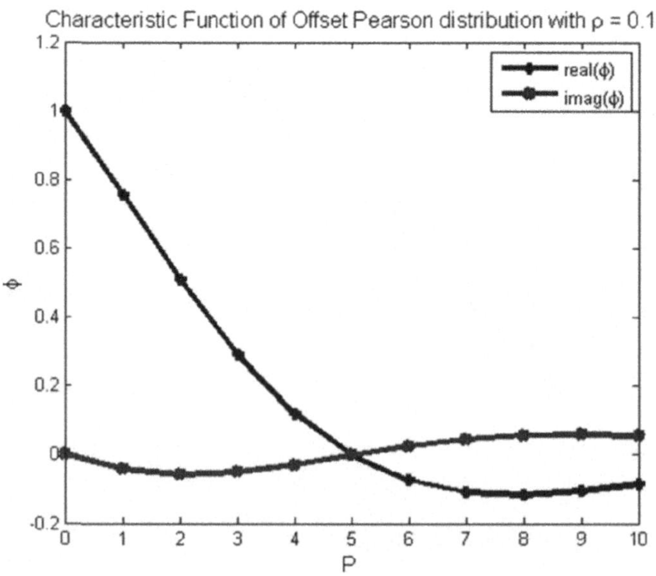

FIGURE 4.5
The characteristic function of offset Pearson distribution with $P = 0.1$.

$\int_C f(z)\,dz = 2\pi i$ (sum of residues at the singular points within C) (Chillingworth 1973)

$$= 2\pi i\, e^{ip\mu} \left\{ \frac{-\sqrt{1-\rho^2}}{\pi} \left(\frac{\beta^{\sigma p} + (-\beta)^{\sigma p}}{2\rho\,(\beta^2 - \alpha^2)} \right) \right\}$$

$$\int_C f(z)\, dz = e^{ip\mu} \frac{\beta^{\sigma p} + (-\beta)^{\sigma p}}{2}$$

$$\varphi_p = e^{ip\mu} \frac{\beta^{\sigma p} + (-\beta)^{\sigma p}}{2} \text{ for } p \in \mathbb{Z}, \theta, \mu \in [0, 2\pi) \text{ and } |\rho| < 1 \quad (4.4.4)$$

Required trigonometric moments are computed from the characteristic function. The population characteristics of the Offset Pearson Type II model for various values of ρ are evaluated in MATLAB by employing the expressions from Mardia and Jupp (2000).

From the population characteristics, it can be observed that with increasing value of parameter ρ, the circular variance gradually decreases, the distribution is symmetric and leptokurtic.

4.5 Offset t-Distribution

Bivariate t-distribution (Radhika 2014) arises as a derived sampling distribution from the bivariate normal distribution and the Chi-square distribution (Anderson 2003). Applications of bivariate t-distribution involve the construction of double sample tests for hypothesis pertaining to the mean of a normal distribution with unknown variance (William et al. 1974). The use of such distributions is enjoying renewed interest due to applications in mathematical finance, especially through the use of the student t-copula (Cherubini et al. 2004). The joint density function of bivariate t-distribution is (Balakrishnan and Chin-Diew 2008, p. 352)

$$f(x, y) = \frac{1}{2\pi\sqrt{1-\rho^2}} \left[1 + \frac{(x^2 - 2\rho x y + y^2)}{q(1-\rho^2)} \right]^{-\frac{(q+2)}{2}} \quad (4.5.1)$$

$$q > 0, -1 < \rho < 1, x, y > 0$$

4.5.1 Probability Density Function (pdf) and Cumulative Distribution Function (cdf) of Offset t-Distribution

Applying offsetting on bivariate t-distribution, pdf and cdf of the offset t-distribution are

$$g(\theta) = \int_0^\infty f(r\cos\theta, r\sin\theta)\, r\, dr$$

$$= \frac{1}{2\pi\sqrt{1-\rho^2}} \int_0^\infty \left[1 + \frac{(1-\rho\sin 2\theta)}{q(1-\rho^2)} r^2\right]^{-\left(\frac{q+2}{2}\right)} r\, dr$$

$$= \frac{\sqrt{1-\rho^2}}{2\pi(1-\rho\sin 2\theta)}$$

Clearly, the density functions of the OP-II distribution and the offset t-distribution are the same. Hence, the cdf and the characteristic functions are also the same as the OP-II distribution, therefore, not presented here to avoid being monotonous.

References

Anderson, T.W. 2003. *An Introduction to Multivariate Statistical Analysis*, John Wiley & Sons, New York, p. 289.

Armstrong, N., Kalceff, W., Cline, J., Bonevich, J., Lynch, P., Tang, C. and Thompson, S. 2004. X-ray Diffraction Characterisation of Nanoparticles Size and Shape Distributions: Application to Bimodal Distributions, *Australian Institute of Physics Publications, Sydney*, pp. 1–3. http://hdl.handle.net/10453/1688.

Balakrishnan, N. and Chin–Diew, L.A.I. 2008. *Continuous Bivariate Distributions*, Springer, New York.

Cherubini, U., Luciano, E. and Vecchiato, W. 2004. *Copula Methods in Finance*, John Wiley & Sons, Hoboken, NJ.

Chillingworth, H.R. 1973. *Complex Variables*, Pergamon Press, New York.

Girija, S.V.S., Radhika, A.J.V. and Dattatreya Rao, A.V. 2013b. On Bimodal Offset Cauchy Distribution, *Journal of the Applied Mathematics, Statistics and Informatics (JAMSI)*, 9(1), 61–67.

Jammalamadaka, S.R. and Sen Gupta, A. 2001. *Topics in Circular Statistics*, World Scientific Press, Singapore.

Lee, M.L.T. and Gross, A.J. 1991. Lifetime Distributions Under Unknown Environment, *Journal of Statistical Planning and Inference*, 29, 137–143.

Mardia, K.V. and Jupp, P.E. 2000. *Directional Statistics*, John Wiley & Sons, Chichester, UK.

Mason, C.C., Hanson, R. L., Ossowski, V., Bian, L., Baier, L.J., Krakoff, J. and Bogardus, C. 2011. Bimodal Distribution of RNA Expression Levels in Human Skeletal Muscle Tissue, *BMC Genomics*, 12, 98. doi:10.1186/1471-2164-12-98.

Nayak, T.K. 1987. Multivariate Lomax Distribution: Properties and Usefulness in Reliability Theory, *Journal of Applied Probability*, 24, 170–177.

Radhika, A.J.V. 2014. Mathematical Tools in the Construction of New Circular Models, PhD dissertation, Acharya Nagarjuna University, Guntur, India.

Radhika, A.J.V., Girija, S.V.S. and Dattatreya Rao, A.V. 2013a. On Univariate Offset Pearson Type II Model: Application to Live Data, *International Journal of Mathematics and Statistics Studies*, 1(1), 1–9.

Ramabhadra Sarma, I., Dattatreya Rao, A.V. and Girija, S.V.S. 2009. On Characteristic Functions of Wrapped Half Logistic and Binormal Distributions, *International Journal of Statistics and Systems*, 4(1), 33–45.

Ramabhadra Sarma, I., Dattatreya Rao, A.V. and Girija, S.V.S. 2011. On Characteristic Functions of Wrapped Lognormal and Weibull Distributions, *Journal of Statistical Computation and Simulation*, 81(5), 579–589.

Vilela, A.L.M., Moreira, F.G.B. and de Souza, A.J.F. 2012. Majority – Vote Model with a Bimodal Distribution of Noises, http://dx.doi.org/10.1016/j.physa.2012.07.068.

William, G.B., Dykstra, R.L. and Hewett, J.E. 1974. A Bivariate t-Distribution with Applications, *Journal of the American Statistical Association*, 69(346), 525–532.

5

Angular Models with New Techniques

5.1 Introduction to the Rising Sun Circular Models

The available methods of generating circular models are wrapping a linear model, offsetting a bivariate linear model and applying stereographic projection on a linear model. The Rising Sun function (Van Rooij and Schikof 1982, p. 10) smoothens the existing curve, and many bumps disappear. This may lead to the effect of increasing the smoothing the curve in density estimation. Girija (2010) proposed a new method of generating circular model by using the Rising Sun function (RSF). Motivated by the mathematical significance of the RSF behind the construction of circular models, an attempt is made to construct new circular models, and various properties are discussed.

5.2 Methodology of Constructing the Rising Sun Circular Models (Girija 2010)

The RSF of a bounded function $f:[a,b]\rightarrow \mathbb{R}$ is defined by

$$f_{\Theta}(x)=\sup\{f(t):x\leq t\leq b\} \tag{5.2.1}$$

It is easy to show that

- when f is nonnegative, then f_{Θ} is nonnegative
- when f is continuous, then f_{Θ} is continuous
- f_{Θ} is monotonically decreasing, hence $f_{\Theta}=f$ when f is decreasing and f_{Θ} is the smallest monotonically decreasing function such that $f_{\Theta}=f$.

Imagine the Rising Sun on the *x*-axis. Then $\{(x,y)\in\mathbb{R}^2 : y \geq f_\Theta(x)\}$ is illuminated by the sun whereas $\{(x,y)\in\mathbb{R}^2 : y < f_\Theta(x)\}$ is covered by darkness. The set $\{(x, f(x)) : f(x) = f_\Theta(x)\}$ is the collection of those points of the graph of *f* that receive light from the sun.

Theorem 5.2.1: (Van Rooij and Schikof 1982)

If *f* is continuous on [*a*, *b*] then for any *k* in the range of f_Θ, $S = \{x \in [a,b] : f_\Theta(x) = k\}$ is a closed and bounded interval.

Proof: Given *f* is continuous on [*a*, *b*] and

$$S = \{x \in [a,b] : f_\Theta(x) = k\}.$$

$$S \subset [a,b] \Rightarrow S \text{ is bounded.}$$

$\{k\}$ is in the range of f_Θ and $\{k\}$ is closed

$$\Rightarrow f_\Theta^{-1}\{k\} \text{ is closed in} [a,b]$$

$$\Rightarrow S \text{ is closed.}$$

Let $\sup S = \beta$ and $\inf S = \alpha$.

$\alpha, \beta \in S.$

Consider $$\alpha \leq x \leq \beta \Rightarrow f_\Theta(\beta) \leq f_\Theta(x) \leq f_\Theta(\alpha)$$

$$\Rightarrow k \leq f_\Theta(x) \leq k$$

$$\Rightarrow f_\Theta(x) = k$$

$$\Rightarrow x \in S$$

$$x < \alpha \Rightarrow f_\Theta(x) \geq f_\Theta(\alpha) = k$$

$$\Rightarrow f_\Theta(x) \geq k$$

If $f_\Theta(x) = k$, then α is not $\inf S$

Hence $f_\Theta(x) \neq k$.

$$x < \alpha \Rightarrow x \notin S \text{ and } x > \beta \Rightarrow x \notin S.$$

$$S = [\alpha, \beta].$$

The Rising Sun Lemma 5.2.2: (Van Rooij and Schikof 1982)

Let $f:[a,b]\rightarrow\mathbb{R}$ be a continuous function. Then $E=\{x\in(a,b): f_{\Theta}(x)>f(x)\}$ is open. If (α,β) is a component of E, then $f_{\Theta}(\beta)=f(\beta)$ and $f_{\Theta}(\alpha)=f(\alpha)$ for $\alpha\neq a$, and f_{Θ} is constant in that interval.
The well-known Lebesgue Theorem is also proved using the RSF.

A new construction procedure of a class of circular models using RSF is obtained in the following theorem, and an illustration is also included in this section. These distributions are named "Rising Sun circular models."

Theorem 5.2.3: (Girija 2010)

If g is the pdf and G is the cdf of a random variable of a circular distribution, then the RSF g_{Θ}, gives rise to the pdf g_c of a circular model. The distribution function of g_c is given by

$$G_c=\begin{cases}\frac{1}{K}\left[\theta_1\, g(\theta_1)+G(\theta)-G(\theta_1)\right] & \text{for } \theta_1<\theta\\[2ex] \frac{1}{K}\left[\theta\, g(\theta_1)\right] & \text{for } \theta_1\geq\theta\end{cases}\tag{5.2.2}$$

Proof: If g is the pdf of r.v. θ of a circular distribution, the RSF g_{Θ} of g is given by

$$g_{\Theta}(\theta)=\sup_{\theta\leq t<2\pi}\{g(t)\}\quad\forall\quad\theta\in[0,\,2\pi)\tag{5.2.3}$$

The function g_{Θ} is normalized by

$$g_c(\theta)=\frac{g_{\Theta}(\theta)}{\int_0^{2\pi} g_{\Theta}(\theta)\ d\theta}\tag{5.2.4}$$

The integral in the denominator is positive since $g_{\Theta}(\theta)\geq g(\theta)$, $\forall\theta$, and $\int_0^{2\pi} g(\theta)d\theta=1$. g_c is extended to $\mathbb{R}$ by $g_c(\theta+2k\pi)=g_c(\theta)$ for every integer k and $\theta\in[0,\,2\pi)$.

Clearly, $g_c(\theta)\geq 0\ \forall\ \theta\in[0,2\pi)$, and g_c is **periodic**.

Since g is continuous and $g(\theta)>0\ \forall\ \theta\in[0,2\pi)$,

$$K=\int_0^{2\pi} g_{\Theta}(\theta)\ d\theta>0$$

Hence $\int_0^{2\pi} g_c(\theta)\ d\theta \ = 1.$

Hence g_c is the pdf of a circular model.

Let G_c be the cdf of the Rising Sun circular model. Suppose the continuous function g has maximum at $\theta = \theta_1$. Then g is monotonically increasing for $0 \le \theta \le \theta_1$ and g is monotonically decreasing for $\theta_1 < \theta < 2\pi$.

Hence

$$g_\Theta(\theta) = \begin{cases} g(\theta_1) & \text{for } 0 \le \theta \le \theta_1 \\ g(\theta) & \text{for } \theta_1 < \theta < 2\pi \end{cases} \tag{5.2.5}$$

The cdf of the Rising Sun circular model is

$$\begin{aligned} G_c(\theta) &= \int_0^{\theta} g_c(\theta')\ d\theta' \\ &= \frac{1}{K}\int_0^{\theta} g_\Theta(\theta')\ d\theta' \end{aligned} \tag{5.2.6}$$

Case (i) Suppose $\theta_1 < \theta$

$$\begin{aligned} G_c(\theta) &= \frac{1}{K}\int_0^{\theta_1} g_\Theta(\theta')\ d\theta' + \frac{1}{K}\int_{\theta_1}^{\theta} g_\Theta(\theta')\ d\theta' \\ &= \frac{1}{K}\int_0^{\theta_1} g(\theta_1)\ d\theta' + \frac{1}{K}\int_{\theta_1}^{\theta} g(\theta')\ d\theta' \\ &= \frac{1}{K}\left[\theta_1\, g(\theta_1) + G(\theta) - G(\theta_1)\right] \end{aligned}$$

Case (ii) Suppose $\theta_1 \ge \theta$

$$\begin{aligned} G_c(\theta) &= \int_0^{\theta} g_c(\theta')\ d\theta' \\ &= \frac{1}{K}\int_0^{\theta} g_\Theta(\theta')\ d\theta' \\ &= \frac{1}{K}\int_0^{\theta} g(\theta_1)\ d\theta' \\ &= \frac{\theta\ g(\theta_1)}{K} \end{aligned}$$

Analogous to Theorem 5.2.2, the Rising Sun lemma for circular data is presented as follows:

The Circular Rising Sun Lemma 5.2.4:

Let $g:[0,2\pi]\to\mathbb{R}$ be a continuous function. Then $E=\{\theta\in(0,2\pi): g_\Theta(\theta)>g(\theta)\}$ is open. If (α,β) is a component of E, then $g_\Theta(\beta)=g(\beta)$ and $g_\Theta(\alpha)=g(\alpha)$ for $\alpha\neq 0$, and g_Θ is constant in that interval, and g_c is the normalized function of g_Θ; hence, β represents the mode of the circular model at which both g_Θ and g_c have maximum value. Hence, a component of g_c as well as g_Θ is $(0,\beta)$ and g_c is the pdf of circular model known as the Rising Sun circular model.

The characteristic function of the Rising Sun circular model is derived here under:

Theorem 5.2.5

Let g and φ_p be the pdf and the characteristic functions of a circular model and θ_1 be the mode of g. Then the characteristic function of the corresponding Rising Sun circular model with pdf g_c is

$$\varphi_\Theta(p)=\begin{cases} 1 \text{ for } p=0 \\ \dfrac{\varphi_p(\theta)}{k}+\dfrac{g(\theta_1)}{kp}i\left(1-e^{ip\theta_1}\right)-\dfrac{1}{k}\displaystyle\int_0^{\theta_1} e^{ip\theta}\,g(\theta)\,d\theta \text{ for } p\neq 0 \end{cases} \tag{5.2.7}$$

where g_Θ is the RSF of g, and $k=\displaystyle\int_0^{2\pi} g_\Theta(\theta)\ d\theta$.

Proof: The pdf of the Rising Sun circular model is

$$g_c(\theta)=\frac{1}{k}g_\Theta(\theta) \tag{5.2.8}$$

For $p\neq 0$

The characteristic function of the Rising Sun circular model is

$$\varphi_\Theta(p)=\int_0^{2\pi} e^{ip\theta}g_c(\theta)d\theta$$

$$=\frac{1}{k}\int_0^{2\pi} e^{ip\theta}g_\Theta(\theta)d\theta$$

$$\varphi_\Theta(p) = \frac{1}{k}\left[\int_0^{\theta_1} e^{ip\theta} g_\Theta(\theta)d\theta + \int_{\theta_1}^{2\pi} e^{ip\theta} g_\Theta(\theta)d\theta\right]$$

$$= \frac{1}{k}\left[\int_0^{\theta_1} e^{ip\theta} g(\theta_1)d\theta + \int_{\theta_1}^{2\pi} e^{ip\theta} g(\theta)d\theta\right]$$

$$= \frac{g(\theta_1)}{k}\left(\frac{-i\,e^{ip\theta}}{p}\right)_0^{\theta_1} + \frac{1}{k}\phi_p(\theta) - \frac{1}{k}\int_0^{\theta_1} e^{ip\theta} g(\theta)d\theta$$

by Theorem 5.2.3 $g_\Theta(\theta) = g(\theta_1)$ for $\theta \le \theta_1$

$= g(\theta)$ for $\theta_1 < \theta$

$$\text{for}\, p = 0,\ \varphi_\Theta(0) = \frac{1}{k}\int_0^{2\pi} g_\Theta(\theta)\ d\theta = 1$$

$$\text{Hence } \varphi_\Theta(p) = \begin{cases} 1 \text{ for } p = 0 \\ \dfrac{\phi_p(\theta)}{k} + \dfrac{g(\theta_1)}{kp} i\left(1 - e^{ip\theta_1}\right) - \dfrac{1}{k}\displaystyle\int_0^{\theta_1} e^{ip\theta} g(\theta)d\theta \ \text{ for}\, p \ne 0 \end{cases} \tag{5.2.9}$$

5.3 Rising Sun von Mises Model

The von Mises distribution was introduced by von Mises (1918) in order to study the deviations of measured atomic weights from integral values. A technique is proposed in which each of these microphone-pair determined azimuths are further combined into a mixture of the von Mises distributions, thus producing a practical probabilistic representation of the microphone array measurement (Radhika et al. 2013). It is shown that this distribution is inherently multimodal and that the system at hand is non-linear, which requires a discrete representation of the distribution function by means of particle filtering (Marković and Petrovic 2010). A multivariate extension of the bivariate model of Singh et al. (2002) is proposed by Mardia et al. (2008). Procedures for the estimation of parameters of the proposed distribution include the method of moments and pseudolikelihood; the efficiency of the latter is investigated in two and three dimensions. The methods are applied to real protein data of conformational angles.

5.3.1 Probability Density Function (pdf) of the Rising Sun von Mises Distribution (RSVMD)

The pdf of the von Mises distribution (Jammalamadaka and Sen Gupta 2001) is

$$g(\theta)=\frac{1}{2\pi I_0(k)}\exp\left(k\cos(\theta-\mu)\right),\ k>0,\ \mu\in[0,2\pi) \tag{5.3.1}$$

where I_0 denotes the modified Bessel function of the first kind and order zero, which can be defined by

$$I_0(k)=\frac{1}{2\pi}\int_0^{2\pi} e^{k\cos\theta}\,d\theta.$$

The function I_0 has power series expansion

$$I_0(k)=\sum_{r=0}^{\infty}\frac{1}{(r!)^2}\left(\frac{k}{2}\right)^{2r}$$

The parameter μ is the mean direction, and the parameter k is known as the concentration parameter. The mean resultant length ρ is $A(k)$, where A is the function defined by

$$A(k)=\frac{I_1(k)}{I_0(k)}$$

The RSF of the von Mises distribution is given by

$$\begin{aligned} g_{\Theta}(\theta) &= \mathrm{Sup}\left(g(t):\theta\le t<2\pi\right) \\ &= \mathrm{Sup}\left(\frac{1}{2\pi I_0(k)}\exp\left(k\cos(t-\mu)\right):\theta\le t<2\pi\right) \end{aligned} \tag{5.3.2}$$

Normalizing this function with the constant $K_1=\int_0^{2\pi} g_{\Theta}(\theta)\ d\theta$ the pdf of the RSVMD is obtained.

$$g_c(\theta)=\frac{\mathrm{Sup}\left(\frac{1}{2\pi I_0(k)}\exp\left(k\cos(t-\mu)\right):\theta\le t<2\pi\right)}{\int_0^{2\pi} g_{\Theta}(\theta)\ d\theta} \tag{5.3.3}$$

The graphs of pdf of the von Mises, Rising Sun of von Mises, and pdf of Rising Sun von Mises models are plotted in Figure 5.1.

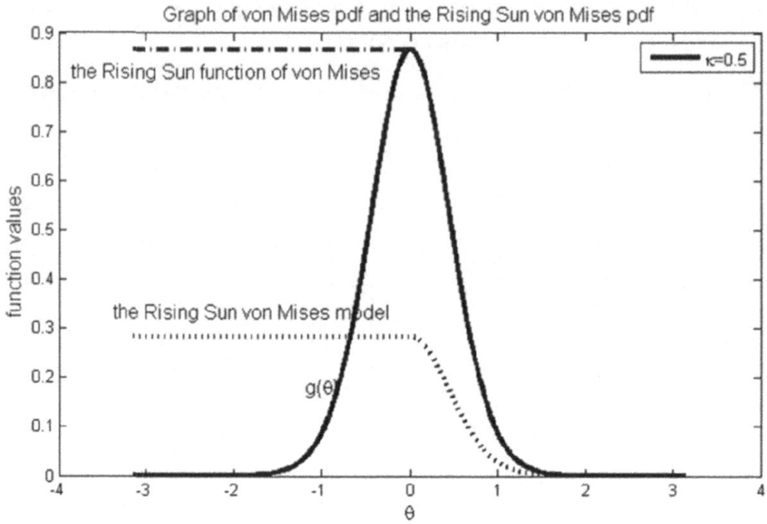

FIGURE 5.1
Graph of von Mises pdf and the Rising Sun von Mises pdf.

5.3.2 The Characteristic Function and the Population Characteristics of the Rising Sun von Mises Model

Since the RSVMD is symmetrical about μ (Figure 5.2),

$$\beta_p = E\left[\sin p(\theta - \mu)\right]$$

$$\alpha_p = \frac{1}{2\pi I_0(k)} \int_0^{2\pi} \cos p(\theta - \mu) e^{k\cos(\theta-\mu)}\, d\theta$$

$$= \frac{I_p(k)}{I_0(k)}$$

where I_p is the modified Bessel function of the first kind and order p.

The modified I_p is defined by $I_p(k) = \frac{1}{2\pi}\int_0^{2\pi} \cos p\theta\, e^{k\cos\theta} d\theta$.

To study the characteristics of the circular models, the trigonometric moments, which are the real and imaginary parts of the characteristic function, are required. As the pdf of the newly constructed Rising Sun circular model is not in closed form, the values of the characteristic function can be evaluated using numerical methods in MATLAB®. The characteristics for the Rising Sun von Mises model are also based on their respective trigonometric moments. These can be expressed in terms of trigonometric moments α_p and β_p. From the population characteristics, it can be observed

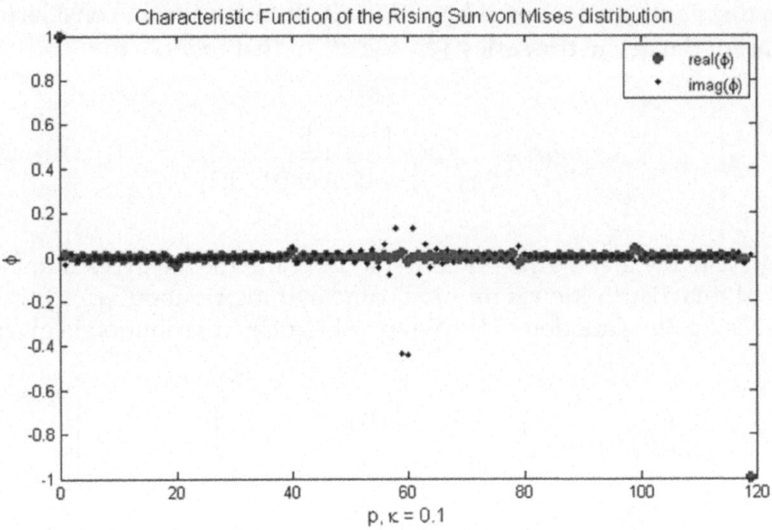

FIGURE 5.2
Graph of the characteristic function of the Rising Sun von Mises distribution.

that with increasing value of parameter *k*, the circular variance gradually decreases, and the distribution is negatively skewed and platykurtic.

5.4 Rising Sun Wrapped Cauchy Distribution (RSWCD) (Radhika et al. 2013)

The Wrapped Cauchy distribution was introduced by Levy (1939) and has been studied by Wintner (1947). McCullagh (1996) showed that the wrapped Cauchy distribution can be obtained by mapping a Cauchy distribution onto the circle by the transformation $x \leftrightarrow 2\tan^{-1} x$.

5.4.1 Probability Density Function (pdf) of the Rising Sun Wrapped Cauchy Distribution

The pdf of the wrapped Cauchy distribution (Mardia and Jupp 2000) is

$$g(\theta) = \left(\frac{1}{2\pi} \left(1 + 2\sum_{p=1}^{\infty} \rho^p \cos p(\theta - \mu) \right) \right) \tag{5.4.1}$$

where $\rho = \pi e^{-a}, a > 0$, and $\theta \in [0, 2\pi)$.

This distribution is denoted by WC (μ, ρ). It follows from considering the real part of the geometric series $\sum_{p=1}^{\infty} \rho^p e^{-ip(\theta-\mu)}$ that reduces to

$$g(\theta) = \frac{\left(1-\rho^2\right)}{2\pi\left(1+\rho^2-2\rho\cos(\theta-\mu)\right)} \tag{5.4.2}$$

In particular, the mean direction is $\mu(\text{mod}\,2\pi)$ and the mean resultant length is ρ, and this distribution is unimodal and symmetric about μ.

The Rising Sun function of the wrapped Cauchy distribution is given by

$$g_{\Theta}(\theta) = \sup\left(\frac{1}{2\pi}\left(\frac{\left(1-\rho^2\right)}{\left(1+\rho^2-2\rho\cos(t-\mu)\right)}\right) : \theta \le t < 2\pi\right) \tag{5.4.3}$$

Normalizing this function with the constant $K_2 = \int_0^{2\pi} g_{\Theta}(\theta)\ d\theta$ the pdf of the RSWCD is obtained.

$$g_c(\theta) = \frac{\sup\left(\frac{1}{2\pi}\left(\frac{\left(1-\rho^2\right)}{\left(1+\rho^2-2\rho\cos(t-\mu)\right)}\right) : \theta \le t < 2\pi\right)}{\int_0^{2\pi} g_{\Theta}(\theta)\ d\theta} \tag{5.4.4}$$

The graphs of pdf of the wrapped Cauchy, Rising Sun of wrapped Cauchy, and pdf of Rising Sun wrapped Cauchy models are plotted (Figure 5.3).

5.4.2 The Characteristic Function and the Population Characteristics of the Rising Sun Wrapped Cauchy Model

As explained earlier in 5.3.2, the characteristics for the Rising Sun wrapped Cauchy models are also based on their respective trigonometric moments. These can be expressed in terms of trigonometric moments α_p and β_p (Figure 5.4).

From the population characteristics, it can be observed that with increasing value of parameter ρ, the circular variance gradually decreases, and the distribution is negatively skewed and platykurtic.

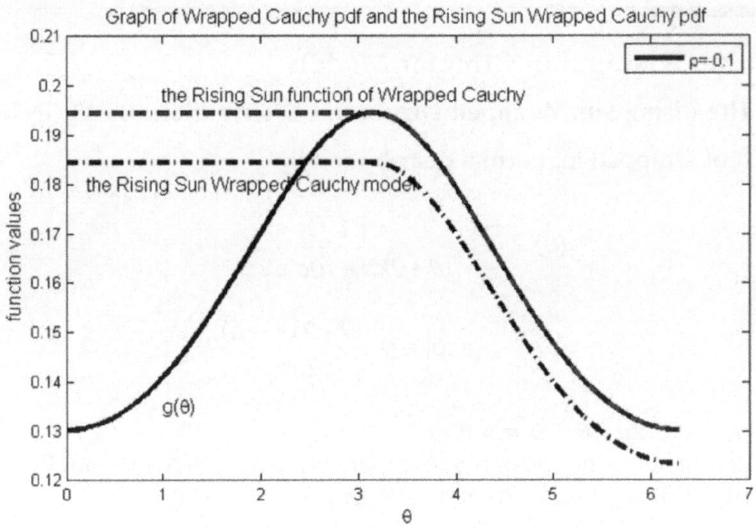

FIGURE 5.3
Graph of the wrapped Cauchy pdf and the Rising Sun wrapped Cauchy pdf.

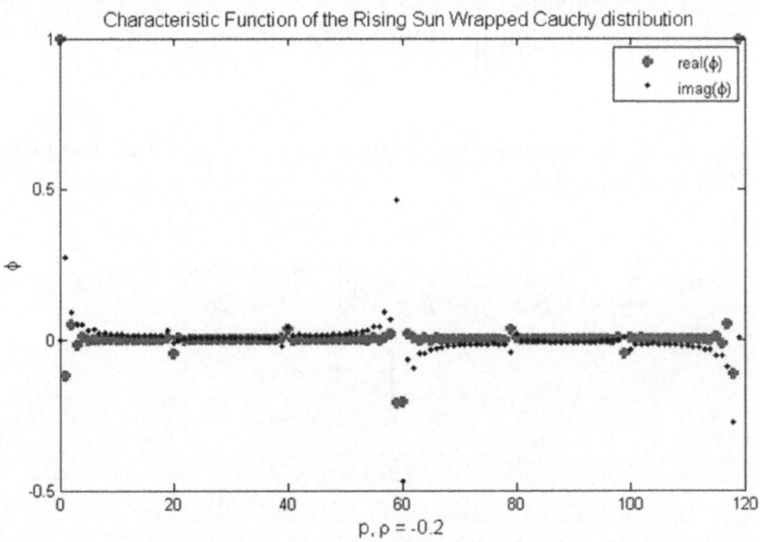

FIGURE 5.4
Graph of the characteristic function of the Rising Sun wrapped Cauchy distribution.

5.5 Other Rising Sun Circular Models

5.5.1 The Rising Sun Wrapped Lognormal Distribution (RSWLGND)

The pdf of wrapped lognormal distribution is

$$g(\theta)=\sum_{k=0}^{\infty}\frac{1}{(\theta+2k\pi-\mu)\sigma\sqrt{2\pi}}\exp\left(-\frac{\{\ln(\theta+2k\pi-\mu)\}^2}{2\sigma^2}\right), \tag{5.5.1}$$

where $\theta,\mu\in[0,2\pi),\theta>\mu,\sigma>0$.

The Rising Sun function of the wrapped lognormal distribution (Dattatreya Rao et al. 2013) is given by for $\theta,\mu\in[0,2\pi),\theta>\mu,\sigma>0$.

$$g_{\Theta}(\theta)=Sup\big(g(t):\theta\le t<2\pi\big)$$
$$=Sup\left(\sum_{k=0}^{\infty}\frac{1}{(t+2k\pi-\mu)\sigma\sqrt{2\pi}}\exp\left(-\frac{\{\ln(\theta+2k\pi-\mu)\}^2}{2\sigma^2}\right):\theta\le t<2\pi\right) \tag{5.5.2}$$

Normalizing this function with the constant $K=\int_0^{2\pi}g_{\Theta}(\theta)\,d\theta$ the pdf of the RSWLGND is obtained.

$$g_c(\theta)=\frac{\text{Sup}\left(\sum_{k=0}^{\infty}\frac{1}{(t+2k\pi-\mu)\sigma\sqrt{2\pi}}\exp\left(-\frac{\{\ln(\theta+2k\pi-\mu)\}^2}{2\sigma^2}\right):\theta\le t<2\pi\right)}{\int_0^{2\pi}g_{\Theta}(\theta)\,d\theta} \tag{5.5.3}$$

5.5.2 The Rising Sun Wrapped Exponential Distribution

The pdf of wrapped exponential distribution is

$$g(\theta)=\frac{\lambda e^{-\lambda\theta}}{1-e^{-2\pi\lambda}}\quad \text{where } \lambda>0 \text{ and } 0\le\theta<2\pi \tag{5.5.4}$$

The RSF of the exponential distribution is given by

$$g_{\Theta}(\theta) = \sup\left(\frac{\lambda e^{-\lambda t}}{1-e^{-2\pi\lambda}} : \theta \le t < 2\pi\right) \tag{5.5.5}$$

where $\lambda > 0$ and $0 \le \theta < 2\pi$.

The pdf of the Rising Sun wrapped exponential distribution (RSWEXPD) is obtained.

$$g_c(\theta) = \sup\left(\frac{\lambda e^{-\lambda t}}{1-e^{-2\pi\lambda}} : \theta \le t < 2\pi\right) \Bigg/ \int_0^{2\pi} g_{\Theta}(\theta)\, d\theta \tag{5.5.6}$$

where $\lambda > 0$ and $0 \le \theta < 2\pi$.

5.6 Circular Models Using Positive Definite Sequences

Every Toeplitz Hermitian positive definite (THPD) matrix can be associated with a circular model through its characteristic function (Hurwitz 1903; Mardia 1972). The idea of the THPD matrix, a special case of Toeplitz matrix, has a natural extension to an infinite case as well (Devaraaj 2012).

The association between circular models and **"POSITIVE DEFINITE SEQUENCES,"** which is the motivation for generating new methods of constructing circular models for possible future applications/extensions, is discussed. New methods of generating circular models using positive definite sequences are derived.

5.7 Methodology of Construction of Circular Models through Positive Definite Sequences

If φ is the characteristic function of a circular distribution and $\varphi_p = \varphi(p)$, then it is known that

1. $\varphi_0 = 1$
2. $\varphi_{-p} = \bar{\varphi}_p$ for $p \in \mathbb{Z}$ and
3. $\sum_{p=-\infty}^{\infty} \left|\varphi_p\right|^2 < \infty$

The third condition is equivalent to $\sum_{p=0}^{\infty}|\varphi_p|^2 < \infty$.

Further, it is assumed that $\forall\ p \in \mathbb{Z}, |\varphi_p|^2 \le 1$.

If $\varphi_p = \alpha_p + i\beta_p$ where α_p and β_p are real then $\alpha_{-p} = \alpha_p$ and $\beta_{-p} = -\beta_p$.

If $\sum_{p=1}^{\infty}(\alpha_p^2 + \beta_p^2)$ is convergent, the random variable θ has a density that is defined **almost everywhere** by

$$f(\theta) = \frac{1}{2\pi}\sum_{p=-\infty}^{\infty}\varphi_p e^{-ip\theta}$$
$$= \frac{1}{2\pi}\left[1 + 2\sum_{p=1}^{\infty}(\alpha_p \cos p\theta + \beta_p \sin p\theta)\right] \tag{5.7.1}$$

If f is a function of bounded variation, then by Jordan's test the above series (5.7.1) converges, provided f has a finite number of maxima and minima and a finite number of discontinuities in $[0, 2\pi)$

OR

$$f(\theta) = \frac{1}{2}\{f(\theta+0) + f(\theta-0)\}.$$

Thus, the characteristic function φ of a circular distribution is identified by the corresponding sequence $\{\varphi_p\}$. If $a_{ij} = \varphi_{i-j}, i \ge 0, j \ge 0$, then $a_{ij} = \bar{a}_{ji}$. Further, if $A = (a_{ij})$ and $A_m = (a_{ij})\ 1 \le i, j \le m$, then all the leading principal minors of A_m are positive. Thus, by the inversion theorem for the characteristic function, it follows that the matrix A corresponding to a circular model is a THPD matrix.

If $\alpha_p = \beta_p = 0$ for $p > n$, then

$$f(\theta) = \frac{1}{2\pi}\left[1 + 2\sum_{p=1}^{n}(\alpha_p \cos p\theta + \beta_p \sin p\theta)\right]$$

and the corresponding matrix A is given by

$$A = \begin{pmatrix} A_n & O' \\ O & 0 \end{pmatrix}$$

where $A_n = \begin{pmatrix} \varphi_0 & \bar{\varphi}_1 & \bar{\varphi}_2 & \cdots & \bar{\varphi}_{n-1} \\ \varphi_1 & \varphi_0 & \bar{\varphi}_1 & \cdots & \bar{\varphi}_{n-2} \\ \varphi_2 & \varphi_1 & \varphi_0 & \cdots & \bar{\varphi}_{n-3} \\ \cdots & \cdots & \cdots & \cdots & \cdots \\ \varphi_{n-1} & \varphi_{n-2} & \varphi_{n-3} & \cdots & \varphi_0 \end{pmatrix}$ is a THPD matrix.

If m is any positive integer, then
$E\left(\left|\sum_{p=1}^{m} e^{ip\theta} z_p\right|^2\right) \geq 0$ for all complex $z_1, z_2, \ldots z_m$ so that

$$\sum_{p,q=1}^{m} \varphi_{p-q} z_p \bar{z}_q \geq 0.$$

Consequently, the sequence $\{\varphi_p\}$ is positive definite.

The converse is also true. That is, a positive definite sequence $\{\varphi_p\}$ with $\varphi_0 = 1$ represents the Fourier coefficients of a circular distribution.

Motivated by the above observation methods, generating a THPD matrix from certain properties of matrices like a Hadamard product and convex combination of matrices are presented.

Let φ be the characteristic function of a circular model and A be the infinite matrix

$$A = \begin{pmatrix} \varphi_0 & \bar{\varphi}_1 & \bar{\varphi}_2 & \cdots & \bar{\varphi}_{n-1} & \cdots \\ \varphi_1 & \varphi_0 & \bar{\varphi}_1 & \cdots & \bar{\varphi}_{n-2} & \cdots \\ \varphi_2 & \varphi_1 & \varphi_0 & \cdots & \bar{\varphi}_{n-3} & \cdots \\ \cdots & \cdots & \cdots & \cdots & \cdots & \cdots \\ \varphi_{n-1} & \varphi_{n-2} & \varphi_{n-3} & \cdots & \varphi_0 & \cdots \\ \cdots & \cdots & \cdots & \cdots & \cdots & \cdots \end{pmatrix}$$

and A_n be the nth order principal minor of A. It is clear that for every n, A_n is a THPD matrix. Clearly

$$A_{n+1} = \begin{pmatrix} A_n & x_{-n} \\ x_n & \varphi_0 \end{pmatrix} \text{ where } x_n = (\varphi_{n-1}, \varphi_{n-2}, \ldots \varphi_0),\ x_{-n} = (\varphi_{n-1}, \varphi_{n-2}, \ldots \varphi_0)'$$

Proposition 5.7.1

If $\{\varphi_p\}$ and $\{\psi_p\}$ are positive definite sequences, then

1. for $0 \leq \lambda \leq 1$, $\{\lambda\varphi_p + (1-\lambda)\psi_p\}$ is a positive definite sequence.
2. $\{\varphi_p \psi_p\}$ is a positive definite sequence.

Proof

1. For $n > 0$,

$$\text{Let } A_n = \begin{pmatrix} \varphi_0 & \bar{\varphi}_1 & \bar{\varphi}_2 & \cdots & \bar{\varphi}_{n-1} \\ \varphi_1 & \varphi_0 & \bar{\varphi}_1 & \cdots & \bar{\varphi}_{n-2} \\ \varphi_2 & \varphi_1 & \varphi_0 & \cdots & \bar{\varphi}_{n-3} \\ \cdots & \cdots & \cdots & \cdots & \cdots \\ \varphi_{n-1} & \varphi_{n-2} & \varphi_{n-3} & \cdots & \varphi_0 \end{pmatrix}$$

$$\text{and } B_n = \begin{pmatrix} \psi_0 & \bar{\psi}_1 & \bar{\psi}_2 & \dots & \bar{\psi}_{n-1} \\ \psi_1 & \psi_0 & \bar{\varphi}_1 & \dots & \bar{\psi}_{n-2} \\ \psi_2 & \psi_1 & \psi_0 & \dots & \bar{\psi}_{n-3} \\ \dots & \dots & \dots & \dots & \dots \\ \psi_{n-1} & \psi_{n-2} & \psi_{n-3} & \dots & \psi_0 \end{pmatrix}$$

Clearly Det $A_n \geq 0$ and Det $B_n \geq 0$ $\forall n$

$$\sum \left\{ \lambda\varphi_{p-q} + (1-\lambda)\psi_{p-q} \right\} z_p \bar{z}_q = \lambda \sum \varphi_{p-q} z_p \bar{z}_q + (1-\lambda) \sum \psi_{p-q} z_p \bar{z}_q \geq 0$$

Therefore $\left\{ \lambda\varphi_p + (1-\lambda)\psi_p \right\}$ is a positive definite sequence.

2. Further, from Bhatia (2007), the Hadamard product $A_n o B_n$ is positive definite $\forall n$.

   Therefore $\left\{ \varphi_p \psi_p \right\}$ is positive definite.

Illustrations:

Consider the sequence $\left\{ \varphi_p \right\}$ when $\varphi_0 = 1$.

1. When $\varphi_w(p) = e^{ip\mu} \frac{\pi\sigma p}{\sinh \pi\sigma p}, p \in \mathbb{Z}$
   $= \frac{\pi\sigma p}{\sinh \pi\sigma p} (\cos p\mu + i \sin p\mu), p \in \mathbb{Z}$ then $\left\{ \varphi_p \right\}$
   is positive definite and is the Fourier series for the wrapped logistic distribution.
2. When $\varphi_w(p) = 2\left(b_n \cos(p\mu) - c_n \sin(p\mu)\right) + 2i\left(b_n \sin(p\mu) + c_n \cos(p\mu)\right)$
   where:

$$b_n = \int_0^a \cos(p\sigma y) \frac{e^{-y}}{\left(1+e^{-y}\right)^2} dy + \sum_{n=1}^{\infty} (-1)^{n-1} \frac{ne^{-na}}{n^2 + p^2\sigma^2} \left(n \cos(p\sigma a) - p\sigma \sin(p\sigma a)\right)$$

$$c_n = \int_0^a \sin(p\sigma y) \frac{e^{-y}}{\left(1+e^{-y}\right)^2} dy + \sum_{n=1}^{\infty} (-1)^{n-1} \frac{ne^{-na}}{n^2 + p^2\sigma^2} \left(n \sin(p\sigma a) + p\sigma \cos(p\sigma a)\right)$$

for

$p \in \mathbb{Z}$, then $\left\{ \varphi_p \right\}$ is positive definite and is the sequence of Fourier coefficients of the wrapped half logistic distribution.

3. When

$$\varphi_W(p) = e^{ip\mu}\left[\left(\frac{\sigma_1}{\sigma_1+\sigma_2}e^{\frac{-p^2\sigma_1{}^2}{2}} + \frac{\sigma_2}{\sigma_1+\sigma_2}e^{\frac{-p^2\sigma_2{}^2}{2}}\right)\right]$$

$$+ie^{ip\mu}\left(\frac{-2}{\sqrt{\pi}}\frac{\sigma_1}{\sigma_1+\sigma_2}e^{\frac{-p^2\sigma_1{}^2}{2}}\sum_{n=1}^{\infty}\frac{\left(\frac{p\sigma_1}{\sqrt{2}}\right)^{2n-1}}{(2n-1)(n-1)!}\right.$$

$$\left.+\frac{2}{\sqrt{\pi}}\frac{\sigma_2}{\sigma_1+\sigma_2}e^{\frac{-p^2\sigma_2{}^2}{2}}\sum_{n=1}^{\infty}\frac{\left(\frac{p\sigma_2}{\sqrt{2}}\right)^{2n-1}}{(2n-1)(n-1)!}\right)$$

for $p \in \mathbb{Z}$

$$\varphi_W(p) = (a\cos p\mu - b\sin p\mu) + i(a\sin p\mu + b\cos p\mu)$$

where $a = \left(\frac{\sigma_1}{\sigma_1+\sigma_2}e^{\frac{-p^2\sigma_1{}^2}{2}} + \frac{\sigma_2}{\sigma_1+\sigma_2}e^{\frac{-p^2\sigma_2{}^2}{2}}\right)$

$$b = \left(\frac{-2}{\sqrt{\pi}}\frac{\sigma_1}{\sigma_1+\sigma_2}e^{\frac{-p^2\sigma_1{}^2}{2}}\sum_{n=1}^{\infty}\frac{\left(\frac{p\sigma_1}{\sqrt{2}}\right)^{2n-1}}{(2n-1)(n-1)!} + \frac{2}{\sqrt{\pi}}\frac{\sigma_2}{\sigma_1+\sigma_2}e^{\frac{-p^2\sigma_2{}^2}{2}}\sum_{n=1}^{\infty}\frac{\left(\frac{p\sigma_2}{\sqrt{2}}\right)^{2n-1}}{(2n-1)(n-1)!}\right)$$

for $p \in \mathbb{Z}$, then $\{\varphi_p\}$ is positive definite and is the sequence of Fourier coefficients for the wrapped binormal distribution.

4. When $|\rho| < \frac{1}{2}, \mu \in [0, 2\pi), \alpha_1 = 2\rho\cos\mu, \beta_1 = \sin\mu, \alpha_p = \beta_p = 0$ for $p \neq 1$ and $\varphi_{-p} = \bar{\varphi}_p$ for $p \geq 0$, then $\{\varphi_p\}$ is positive definite and induces the tridiagonal matrix

$$A = \begin{pmatrix} \varphi_0 & \bar{\varphi}_1 & 0 & \dots & 0 & \dots \\ \varphi_1 & \varphi_0 & \bar{\varphi}_1 & \dots & 0 & \dots \\ 0 & \varphi_1 & \varphi_0 & \dots & 0 & \dots \\ \dots & \dots & \dots & \dots & \dots & \dots \\ 0 & 0 & 0 & \dots & \varphi_0 & \dots \\ \dots & \dots & \dots & \dots & \dots & \dots \end{pmatrix}$$

The pdf of the corresponding circular distribution is given by

$$g(\theta) = \frac{1}{2\pi}(1 + 2\rho\cos(\theta - \mu)) \ , \ \theta \in [0, 2\pi), -\frac{1}{2} < \rho < \frac{1}{2}$$

Clearly this is the Cardioid distribution with parameters μ and ρ.

It is observed that the matrix A is a tridiagonal THPD matrix.

5.8 Discretization of Continuous Circular Models (Srihari et al. 2018)

In mathematics, **discretization** is the process of transforming continuous functions, models, variables, and equations into its discrete counterparts. This process is usually carried out to model a given discrete phenomenon depending on the practical situation. In this chapter, discretization suggested by (Wang and Shimizu 2014) is adopted to construct discrete circular models by transforming existing continuous circular models.

Consider a continuous circular distribution with probability density function (pdf) $f(\theta)$ on the circle to construct a probability mass function $Pr\left(\theta = \frac{2\pi r}{m}\right)$ on a set of the equally spaced points for $r = 0, 1, \ldots, m - 1$ with a fixed integer m equal or greater than 2.

$Pr\left(\theta = \frac{2\pi r}{m}\right) = f\left(\frac{2\pi r}{m}\right) / \sum_{r=0}^{m-1} f\left(\frac{2\pi r}{m}\right)$ is probability mass function (pmf) of respective discrete circular model. $m = 1$ is excluded as it is degenerate.

5.9 Discrete Wrapped Exponential Distribution

The probability density function of wrapped exponential distribution (Jammalamadaka and Kozubowski 2004) is $f(\theta) = \frac{\lambda e^{-\lambda\theta}}{1-e^{-2\pi\lambda}}$ where $\lambda > 0$ is a parameter.

This model is converted into discrete wrapped exponential distribution (DWEXD) by normalizing the probability density function with $\frac{1-e^{\frac{-2\pi\lambda}{m}}}{\lambda}$.

5.9.1 Probability Mass Function

The probability mass function of the DWEXD is (Figure 5.5)

$$Pr\left(\theta=\frac{2\pi r}{m}\right)=\frac{e^{-\lambda\theta}}{1-e^{-2\pi\lambda}}\left(1-e^{\frac{-2\pi\lambda}{m}}\right) \quad \text{where } m\in\mathbb{Z}^{+} \text{ and } r=0,1,2\ldots m-1$$

5.9.2 Cumulative Distribution Function

The distribution function of the DWEXD is

$$F_w(\theta)=\sum_{r=0}^{k}Pr\left(\theta=\frac{2\pi r}{m}\right)$$

$$=\frac{1-e^{\frac{-2\pi\lambda}{m}}}{1-e^{-2\pi\lambda}}\sum_{r=0}^{k}e^{\frac{-2\pi r\lambda}{m}}=\frac{1-e^{\frac{-2\pi\lambda}{m}}}{1-e^{-2\pi\lambda}}\sum_{r=0}^{k}e^{\frac{-2\pi r\lambda}{m}}$$

$$=\frac{1-e^{\frac{-2\pi\lambda(k+1)}{m}}}{1-e^{-2\pi\lambda}}$$

where $k = 0,1,2\ldots\ m-1$

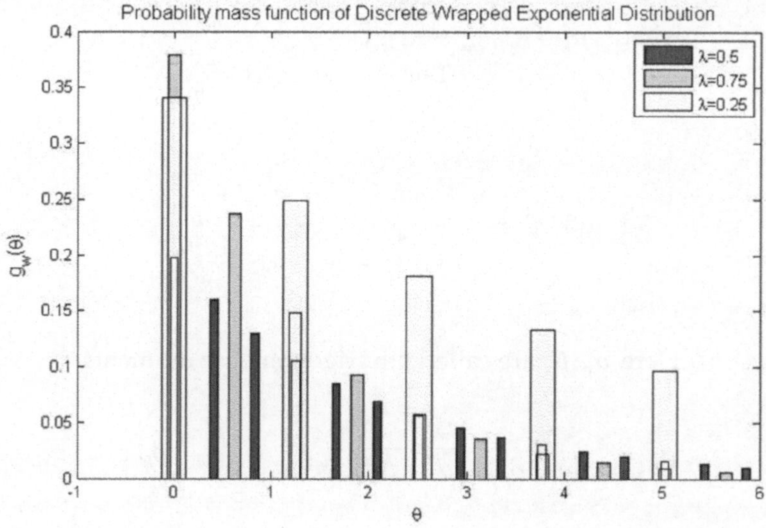

FIGURE 5.5
Probability mass function of discrete wrapped exponential distribution.

5.9.3 Characteristic Function

The characteristic function of the DWEXD is

$$\varphi_\theta(p) = E(e^{ip\theta}) \text{ where } p \in \mathbb{Z}$$

$$= \sum_{r=0}^{m-1} e^{ip\theta} \frac{e^{-\lambda\theta}}{1-e^{-2\pi\lambda}}\left(1-e^{\frac{-2\pi\lambda}{m}}\right)$$

$$= \left(1-e^{\frac{-2\pi\lambda}{m}}\right)\left(\frac{1-e^{\frac{-2\pi\lambda}{m}}\cos\frac{2\pi p}{m} + ie^{\frac{-2\pi\lambda}{m}}\sin\frac{2\pi p}{m}}{1+e^{\frac{-4\pi\lambda}{m}} - 2e^{\frac{-2\pi\lambda}{m}}\cos\frac{2\pi p}{m}}\right)$$

$$= \frac{1-e^{\frac{-2\pi\lambda}{m}}\cos\frac{2\pi p}{m} + ie^{\frac{-2\pi\lambda}{m}}\sin\frac{2\pi p}{m} - e^{\frac{-2\pi\lambda}{m}} + e^{\frac{-4\pi\lambda}{m}}\cos\frac{2\pi p}{m} - ie^{\frac{-4\pi\lambda}{m}}\sin\frac{2\pi p}{m}}{1+e^{\frac{-4\pi\lambda}{m}} - 2e^{\frac{-2\pi\lambda}{m}}\cos\frac{2\pi p}{m}}$$

$$= \alpha_p + i\beta_p$$

$$\text{where } \alpha_p = \frac{1-e^{\frac{-2\pi\lambda}{m}}\cos\frac{2\pi p}{m} - e^{\frac{-2\pi\lambda}{m}} + e^{\frac{-4\pi\lambda}{m}}\cos\frac{2\pi p}{m}}{1+e^{\frac{-4\pi\lambda}{m}} - 2e^{\frac{-2\pi\lambda}{m}}\cos\frac{2\pi p}{m}}$$

$$\text{and } \beta_p = \frac{e^{\frac{-2\pi\lambda}{m}}\sin\frac{2\pi p}{m} - e^{\frac{-4\pi\lambda}{m}}\sin\frac{2\pi p}{m}}{1+e^{\frac{-4\pi\lambda}{m}} - 2e^{\frac{-2\pi\lambda}{m}}\cos\frac{2\pi p}{m}}$$

Here α_p, β_p are called pth trigonometric moments

$$\text{Clearly } \rho_p = \sqrt{\alpha_p^2 + \beta_p{}^2} \text{ and } \mu_p = \tan^{-1}\left(\frac{\beta_p}{\alpha_p}\right)$$

The characteristic function is evaluated and trigonometric moments for the DWEXD are obtained. The population characteristics of the DWEXD are computed using these trigonometric moments (Figure 5.6).

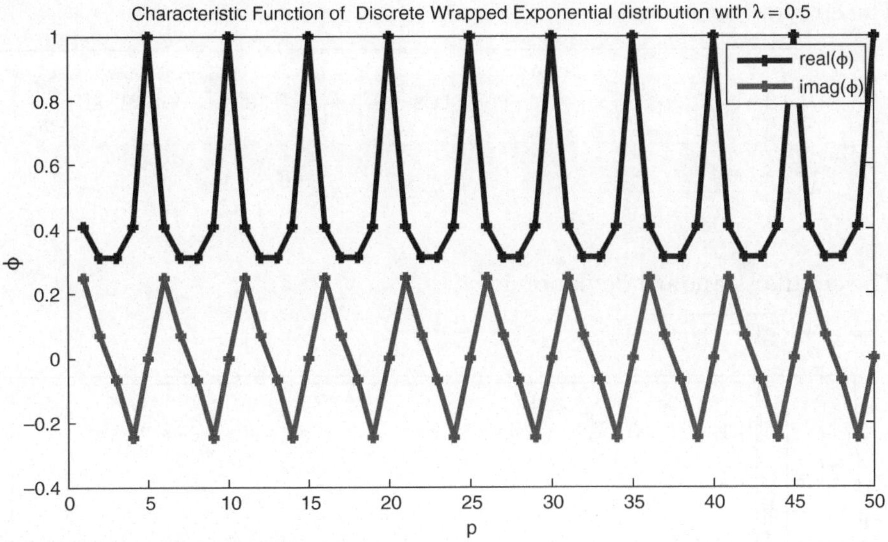

FIGURE 5.6
Characteristic function of discrete wrapped exponential distribution with $\lambda = 0.5$.

5.9.4 Population Characteristics of Discrete Wrapped Exponential Distribution

The population characteristics for the DWEXD for different values of the parameter λ are computed.

The circular mean direction is

$$\mu_1 = \tan^{-1}\left[\frac{e^{\frac{-2\pi\lambda}{m}}\sin\frac{2\pi}{m} - e^{\frac{-4\pi\lambda}{m}}\sin\frac{2\pi}{m}}{1-e^{\frac{-2\pi\lambda}{m}}\cos\frac{2\pi}{m} - e^{\frac{-2\pi\lambda}{m}} + e^{\frac{-4\pi\lambda}{m}}\cos\frac{2\pi}{m}}\right]$$

The concentration is

$$\rho_1 = \sqrt{\left[\frac{\left(1-e^{\frac{-2\pi\lambda}{m}}\cos\frac{2\pi}{m} - e^{\frac{-2\pi\lambda}{m}} + e^{\frac{-4\pi\lambda}{m}}\cos\frac{2\pi}{m}\right)^2 + \left(e^{\frac{-2\pi\lambda}{m}}\sin\frac{2\pi}{m} - e^{\frac{-4\pi\lambda}{m}}\sin\frac{2\pi}{m}\right)^2}{\left(1+e^{\frac{-4\pi\lambda}{m}} - 2e^{\frac{-2\pi\lambda}{m}}\cos\frac{2\pi}{m}\right)^2}\right]}$$

In general $\mu_1 = \mu$ and $\rho_1 = \rho$

The circular variance is

$$V_o = 1 - \sqrt{\left[\frac{\left(1 - e^{\frac{-2\pi\lambda}{m}}\cos\frac{2\pi}{m} - e^{\frac{-2\pi\lambda}{m}} + e^{\frac{-4\pi\lambda}{m}}\cos\frac{2\pi}{m}\right)^2 + \left(e^{\frac{-2\pi\lambda}{m}}\sin\frac{2\pi}{m} - e^{\frac{-4\pi\lambda}{m}}\sin\frac{2\pi}{m}\right)^2}{\left(1 + e^{\frac{-4\pi\lambda}{m}} - 2e^{\frac{-2\pi\lambda}{m}}\cos\frac{2\pi}{m}\right)^2}\right]}$$

The circular standard deviation is

$$\sigma_o = \sqrt{-2\log(1 - V_o)}$$

$$= \sqrt{-2\log\left[\sqrt{\left[\frac{\left(1 - e^{\frac{-2\pi\lambda}{m}}\cos\frac{2\pi}{m} - e^{\frac{-2\pi\lambda}{m}} + e^{\frac{-4\pi\lambda}{m}}\cos\frac{2\pi}{m}\right)^2 + \left(e^{\frac{-2\pi\lambda}{m}}\sin\frac{2\pi}{m} - e^{\frac{-4\pi\lambda}{m}}\sin\frac{2\pi}{m}\right)^2}{\left(1 + e^{\frac{-4\pi\lambda}{m}} - 2e^{\frac{-2\pi\lambda}{m}}\cos\frac{2\pi}{m}\right)^2}\right]}\right]}$$

Central trigonometric moments

The pth central trigonometric moment of θ is

$$\varphi_\theta^*(p) = E\left(e^{ip(\theta-\mu)}\right)$$

$$= E\left(e^{ip\theta}e^{-ip\mu}\right) = e^{-ip\mu}E\left(e^{ip\theta}\right)$$

$$= \alpha_p \cos p\mu + \beta_p \sin p\mu + i(\beta_p \cos p\mu - \alpha_p \sin p\mu)$$

$$= \alpha_p^* + i\beta_p^*$$

where $\alpha_p^* = \alpha_p \cos p\mu + \beta_p \sin p\mu$

$$\beta_p^* = \beta_p \cos p\mu - \alpha_p \sin p\mu$$

The skewness for the DWEXD is

$$\gamma_1 = \frac{\beta_2^*}{(V_o)^{\frac{3}{2}}}$$

$$= \frac{\beta_2 \cos 2\mu - \alpha_2 \sin 2\mu}{\left[1 - \sqrt{\left[\frac{\left(1 - e^{\frac{-2\pi\lambda}{m}}\cos\frac{2\pi}{m} - e^{\frac{-2\pi\lambda}{m}} + e^{\frac{-4\pi\lambda}{m}}\cos\frac{2\pi}{m}\right)^2 + \left(e^{\frac{-2\pi\lambda}{m}}\sin\frac{2\pi}{m} - e^{\frac{-4\pi\lambda}{m}}\sin\frac{2\pi}{m}\right)^2}{\left(1 + e^{\frac{-4\pi\lambda}{m}} - 2e^{\frac{-2\pi\lambda}{m}}\cos\frac{2\pi}{m}\right)^2}\right]}\right]^{\frac{3}{2}}}$$

The kurtosis for the DWEXD is

$$\gamma_2 = \frac{\alpha_2^{*} - (1-V_o)^4}{(V_o)^2}$$

$$= \frac{\left((\alpha_2 \cos 2\mu + \beta_2 \sin 2\mu) - \left[\sqrt{\left[\frac{\left(1 - e^{\frac{-2\pi\lambda}{m}} \cos\frac{2\pi}{m} - e^{\frac{-2\pi\lambda}{m}} + e^{\frac{-4\pi\lambda}{m}} \cos\frac{2\pi}{m}\right)^2 + \left(e^{\frac{-2\pi\lambda}{m}} \sin\frac{2\pi}{m} - e^{\frac{-4\pi\lambda}{m}} \sin\frac{2\pi}{m}\right)^2}{\left(1 + e^{\frac{-4\pi\lambda}{m}} - 2e^{\frac{-2\pi\lambda}{m}} \cos\frac{2\pi}{m}\right)^2} \right]} \right]^4 \right)}{\left[1 - \sqrt{\left[\frac{\left(1 - e^{\frac{-2\pi\lambda}{m}} \cos\frac{2\pi}{m} - e^{\frac{-2\pi\lambda}{m}} + e^{\frac{-4\pi\lambda}{m}} \cos\frac{2\pi}{m}\right)^2 + \left(e^{\frac{-2\pi\lambda}{m}} \sin\frac{2\pi}{m} - e^{\frac{-4\pi\lambda}{m}} \sin\frac{2\pi}{m}\right)^2}{\left(1 + e^{\frac{-4\pi\lambda}{m}} - 2e^{\frac{-2\pi\lambda}{m}} \cos\frac{2\pi}{m}\right)^2} \right]} \right]^2}$$

5.10 Construction of the Circular Model through Differential Equation

It is found that taking suitable initial conditions solution of second-order differential equation is a circular distribution (Dattatreya Rao et al. 2011). For example, considering the differential equation

$$\frac{d^2y}{d\theta^2} + y = \frac{1}{2\pi}, \quad y(0) = \frac{1+2\rho\cos\mu}{2\pi}, \quad y'(0) = \frac{\rho\sin\mu}{\pi} \text{ admits}$$

1. The particular integral, which is a uniform distribution on the unit circle

   and

2. Probability density function of the Cardioid distribution

   $$y(\theta) = \frac{1}{2\pi}\big(1 + 2\rho\cos(\theta-\mu)\big), \text{ where } -\pi \le \theta, \mu < \pi \text{ and } |\rho| < 0.5.$$

   Similarly, taking suitable second-order initial value problems, other new circular models can be constructed.

References

Bhatia, R. 2007. *Positive Definite Matrices*, Princeton University Press, Princeton, NJ.

Dattatreya Rao, A.V., Girija, S.V.S. and Devaraaj, V.J. 2013. On the Rising Sun Wrapped Lognormal and the Rising Sun Wrapped Exponential Models, *International Journal of Statistics and Systems*, 3(1), 1–10.

Dattatreya Rao, A.V., Girija, S.V.S. and Phani, Y. 2011. Differential Approach to Cardioid Distribution, *Computer Engineering and Intelligent Systems*, 2(8), 1–6.

Devaraaj, V. J. 2012. Some Contributions to Circular Statistics, PhD dissertation, Acharya Nagarjuna University, Guntur, India.

Girija, S.V.S. 2010. *Construction of New Circular Models*, VDM Verlag, Saarbrucken, Germany.

Hurwitz, A. 1903. Ueber die Fourierschen Konstanten integierbarer Funktionen, *Mathematische Annalen* **57**, 425–446; 59 553, [86].

Jammalamadaka, S.R. and Kozubowski, T.J. 2004. A Wrapped Exponential Circular Model, *Proceedings of Andhra Pradesh Akademi of Sciences*, India, 43–56.

Jammalamadaka, S.R. and Sen Gupta, A. 2001. *Topics in Circular Statistics*, World Scientific Press, Singapore.

Levy, P. 1939. L'addition des variables aleatories definies sur une circonference, *Bulletin de la Societe Mathematique de France*, 67, 1–41.

Mardia, K.V. 1972. *Statistics of Directional Data*, Academic Press, New York.

Mardia, K. V., Hughes, G. and Taylor, C.C. 2008. A Multivariate von Mises Distribution with Applications to Bioinformatics, *The Canadian Journal of Statistics*, 36(1), 99–109.

Mardia, K.V. and Jupp, P.E. 2000. *Directional Statistics*, John Wiley & Sons, Chicester, UK.

Marković, I. and Petrovic, I. 2010. Applying von Mises Distribution to Microphone Array Probabilistic Sensor Modelling, Robotics (ISR), *41st International Symposium on and 2010 6th German Conference on Robotics (ROBOTIK)*, Munich, Germany.

McCullagh, P. 1996. Mobius Transformation and Cauchy Parameter Estimation, *Annals of Statistics*, 24, 787–808.

Radhika, A.J.V. 2014. Mathematical Tools in the Construction of New Circular Models, PhD dissertation, Acharya Nagarjuna University, Guntur, India.

Radhika, A.J.V., Girija, S.V.S. and Dattatreya Rao, A.V. 2013. On Rising Sun von Mises and Rising Sun Wrapped Cauchy Circular Models, *Journal of the Applied Mathematics, Statistics and Informatics (JAMSI)*, 9(2), 61–67.

Singh, H., Hnizdo, V. and Demchuk, E. 2002. Probabilistic Model for Two Dependent Circular Variables, *Biometrika*, 89, 719–723.

Srihari, G.V.L.N., Girija, S.V.S. and Dattatreya Rao, A.V. 2018. On Discrete Wrapped Exponential Distribution-Characteristics, *International Journal of Scientific Research In Mathematical and Statistical Sciences*, 5(2), 57–64.

Van Rooij, A.C.M. and Schikof, W.H. 1982. *A Second Course on Real Functions*, Cambridge University Press, Cambridge, UK.

von Mises, R. 1918. Uber die "Ganzzahligkeit" der Atomgewichte and verwandte Frogmen, *Physikalische Zeitschrift*, 19, 490–500.

Wang, M.-Z. and Shimizu, K. 2014. Discrete Cardioid Distribution, *Presented in Conference on Advances and Applications in Distribution Theory*, The Institute of Statistical Mathematics, Tokyo, Japan.

Wintner, A. 1947. On the Shape of the Angular Case of Cauchy's Distribution Curves, *Annals of Mathematical Statistics*, 18, 589–593.

6

Extemporaneous Semicircular/Axial Models

6.1 Introduction

It can be observed that Feldspar laths data (Smith 1988) and face-cleat coal seam data (Fisher 1993) spans in the range of $(0,\pi)$ prompting to fit semicircular models for the purpose of data analysis. It is the practice to double the angles to carry out analysis based on circular data and to back transform the results for interpreting semicircular data. Jammalamadaka and Sen Gupta (2001, p. 48) have defined axial distributions restricting a circular model to any arc of arbitrary length, say $\frac{2\pi}{l}$, l is positive integer. Particularly, they defined l-axial circular normal distribution and suggested it to be relevant in modeling axial data of pebbles.

To handle axial data from the data analysis point of view, the approach used hither to transform the axial or l-axial data to vector data as indicated below

$$\theta_i = \begin{cases} 2\theta_i \ (\text{modulo } 360^\circ) & \text{for axial data} \\ l\theta_i \ (\text{modulo } 360^\circ) & \text{for } l\text{-axial data} \end{cases}$$

Then analyze vector data as required and back transform the results. This procedure is suggested mainly because semicircular/axial models are not available in the literature prior to 2001. When such kinds of l-axial models exist, it is natural to use them for the purpose of data analysis. l-axial and l-arc models are used synonymously.

The main objective of this chapter is to construct semicircular and arc models that occur naturally with appropriate construction procedures. It is noticed that inducing inverse stereographic projection on the linear models defined over $(0,\infty)$ results in semicircular models with domain $(-\frac{\pi}{2},\frac{\pi}{2})$ or $(0,\pi)$. Using this idea, stereographic semicircular Weibull distribution (SSCWBD), stereographic semicircular half logistic distribution

(SSCHLD), stereographic semicircular exponentiated inverted Weibull distribution (SSCEIWD), and stereographic semicircular new Weibull–Pareto distribution (SSCNWPD) are constructed on the lines of Minh and Farnum (2003) in this chapter.

Further, it was observed that when offsetting is applied on bivariate beta distribution the resultant model is found to be an arc offset beta distribution, which is a univariate arc model and is presented in this chapter.

Construction of the above angular distributions have occurred as a natural manifestation without imposing any restrictions, hence can be used to model axial data without any modifications or manipulations on the data. Such naturally occurring arc models are called **extemporaneous arc models**.

Having constructed the new offset/stereographic angular models, graphs of pdf, cdf, and characteristic functions of the above distributions are shown. Respective population characteristics are studied through trigonometric moments.

Inverse stereographic projection is also defined by the one-to-one mapping given by $T(\theta)=x=u+v\tan\left(\frac{\theta}{2}\right)$, where $x\in(0,\infty)$, $\theta\in(0,\pi)$, $u\in\mathbb{R}$, and $v>0$.

In the continuous case, $g:[0,\pi)\to\mathbb{R}$ is the probability density function of a semicircular distribution if and only if g has the following basic properties:

$$g(\theta)\geq 0,\quad \forall\theta \tag{6.1.1}$$

$$\int_0^{\pi} g(\theta)\, d\theta = 1 \tag{6.1.2}$$

$$g(\theta)=g(\theta+k\pi),\ k \text{ is an integer} \tag{6.1.3}$$

The characteristic function of a semicircular model with probability density function $g(\theta)$ is defined as $\varphi_p(\theta)=\int_0^{\pi} e^{ip\theta} g(\theta)d\theta$, $p\in\mathbb{Z}$. The characteristic function of a stereographic semicircular model can be obtained in terms of respective linear model.

Theorem 6.1.1

If $G(\theta)$ and $g(\theta)$ are the cdf and the pdf of the stereographic semicircular model and $F(x)$ and $f(x)$ are the cdf and the pdf of the respective linear model, then the characteristic function of stereographic semicircular model is $\varphi_{X_{SC}}(p)=\varphi_{2\tan^{-1}\left(\frac{x}{v}\right)}(p)$, $p\in\mathbb{Z}$.

Proof

$$\varphi_{X_{SC}}(p) = \int_0^{\pi} e^{ip\theta} d(G(\theta)),\ p \in \mathbb{Z}$$

$$= \int_0^{\pi} e^{ip\theta} d\left(F\left(v \tan\frac{\theta}{2}\right)\right)$$

$$= \int_0^{\pi} e^{ip\theta} f\left(v \tan\frac{\theta}{2}\right) d\left(v \tan\frac{\theta}{2}\right)$$

$$= \int_0^{\infty} e^{ip\left(2\tan^{-1}\left(\frac{x}{v}\right)\right)} f(x)dx, \quad \text{putting } x = v\tan\left(\frac{\theta}{2}\right)$$

$$= \varphi_{2\tan^{-1}\left(\frac{x}{v}\right)}(p)$$

As the integral cannot be obtained analytically, Weddle's rule, a numerical technique, is applied for the evaluation of the values of the characteristic function in MATLAB®. The graphs for real and imaginary parts of the characteristic function are plotted.

6.2 Stereographic Semicircular Weibull Distribution (SSCWBD) (Phani et al. 2013)

A random variable X on the real line is said to have the Weibull distribution with shape parameter $c > 0$ and location parameter α if the probability density function and the cumulative distribution function of X are respectively given by

$$f(x) = c(x-\alpha)^{c-1} \exp\left(-(x-\alpha)^c\right) \tag{6.2.1}$$

$$F(x) = 1 - \exp\left(-(x-\alpha)^c\right) \text{ for } x > 0, c > 0 \tag{6.2.2}$$

Then application of inverse stereographic projection defined by a one-to-one mapping $x = v\tan\left(\frac{\theta}{2}\right)$, $v > 0, 0 < \theta < \pi$, leads to the SSCWBD.

6.2.1 Probability Density Function and Cumulative Distribution Function

A random variable X_{SC} on the unit semicircle is said to have the SSCWBD with shape parameter $c > 0$ (Figures 6.1 and 6.2), location parameter μ, and concentration parameter $\lambda > 0$ denoted by SSCWBD (c, λ, μ), if the probability density and the cumulative distribution functions are respectively given by

$$g(\theta)=\frac{\lambda c}{2}\sec^2\left(\frac{\theta}{2}\right)\left(\tan\left(\frac{\theta}{2}\right)-\mu\right)^{c-1}\exp\left(-\lambda\left(\tan\left(\frac{\theta}{2}\right)-\mu\right)^{c}\right), \quad (6.2.3)$$

where $0<\theta<\pi, c>0, \lambda=v^c>0$, and $\mu=\dfrac{\alpha}{v}$.

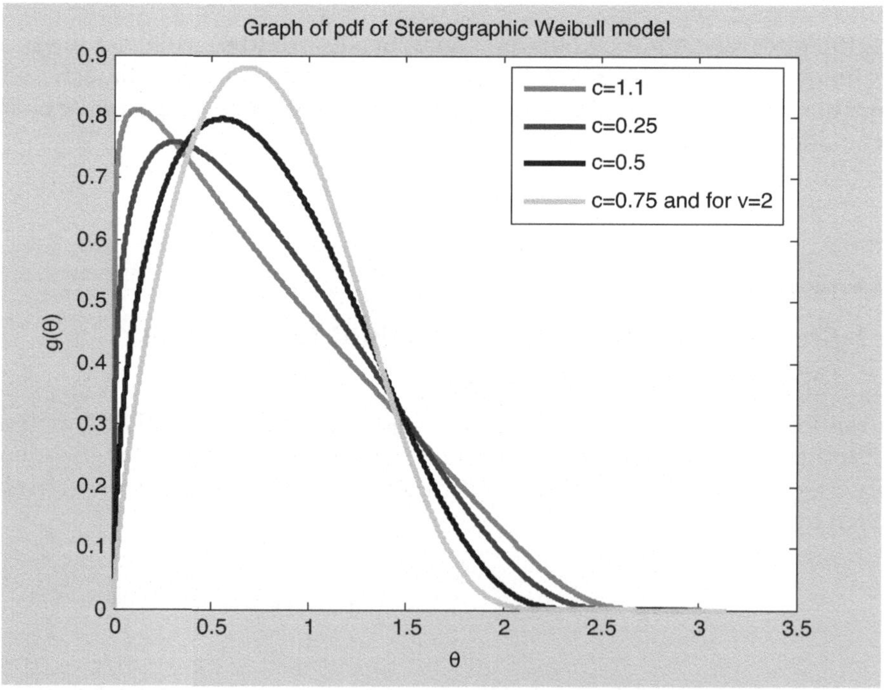

FIGURE 6.1
Graph of the pdf of the stereographic semicircular Weibull distribution for $\mu = 0$.

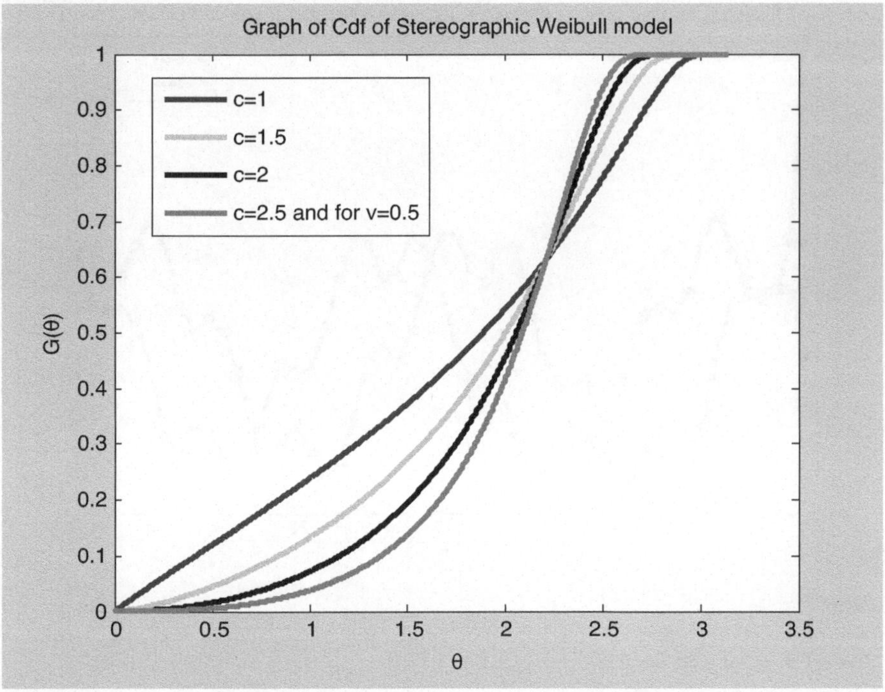

FIGURE 6.2
Graph of the cdf of the stereographic semicircular Weibull distribution for $\mu = 0$.

$$G(\theta) = 1 - \exp\left(-\lambda\left(\tan\left(\frac{\theta}{2}\right) - \mu\right)^{c}\right) \tag{6.2.4}$$

6.2.2 Characteristic Function and Trigonometric Moments

The probability density function of SSCWBD is

$$g(\theta) = \frac{c\lambda}{2}\sec^{2}\left(\frac{\theta}{2}\right)\left(\tan\left(\frac{\theta}{2}\right) - \mu\right)^{c-1} e^{-\lambda\left(\tan\left(\frac{\theta}{2}\right)-\mu\right)^{c}},$$

where $c > 0, \lambda > 0, 0 < \theta < \pi$.

Without loss of generality, it is assumed that $\mu = 0$ in the pdf of SSCWDB model (Figure 6.3).

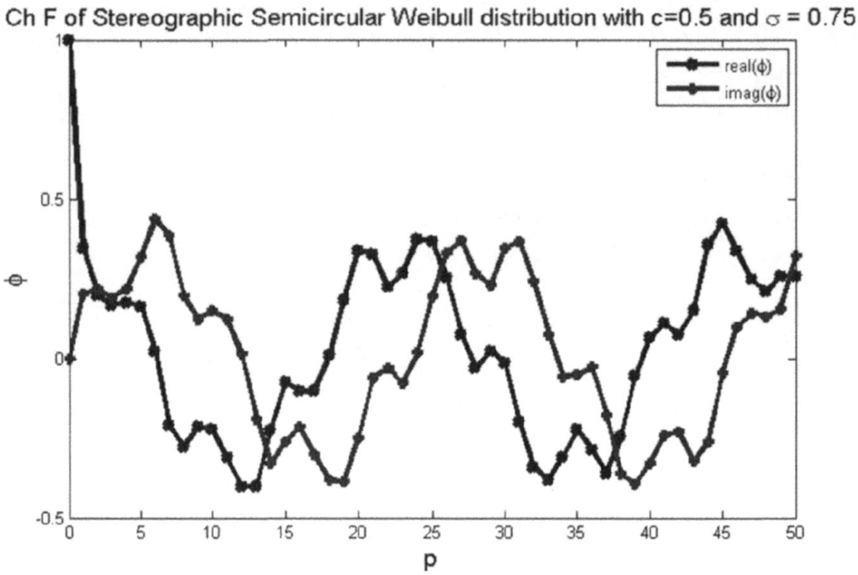

FIGURE 6.3
Graph of the characteristic function of the stereographic semicircular Weibull distribution with $c = 0.5$, $\mu = 0$, and $\lambda = 0.75$.

The characteristic function of SSCWBD model is

$$\varphi_{X_{SC}}(p) = \int_0^{\pi} e^{ip\theta} g(\theta)\, d\theta$$

$$= \frac{\lambda c}{2} \int_0^{\pi} e^{ip\theta} \sec^2\left(\frac{\theta}{2}\right)\left(\tan\left(\frac{\theta}{2}\right)\right)^{c-1} e^{-\lambda\left(\tan\left(\frac{\theta}{2}\right)\right)^{c}} d\theta$$

$$= c \int_0^{\infty} e^{ip\left(2\tan^{-1}(x)\right)} x^{c-1} e^{-x^c} dx$$

The trigonometric moments of the distribution are given by $\{\varphi_p : p = 0, \pm 1, \pm 2, \pm 3, \ldots\}$, where $\varphi_p = \alpha_p + \beta_p$, with $\alpha_p = E(\cos p\theta)$ and $\beta_p = E(\sin p\theta)$ being the pth order cosine and sine moments of the random angle θ, respectively.

Theorem 6.2.1

Under the pdf of the SSCWBD model with $\mu = 0$, the first two $\alpha_p = E(\cos p\theta)$ and $\beta_p = E(\sin p\theta)$, $p = 1, 2$ are given as follows:

$$\alpha_1 = 1 - 2\lambda \sum_{n=0}^{\infty} (-1)^n (\lambda)^{-\left(\frac{2n+c+2}{c}\right)} \Gamma\left(\frac{2n+c+2}{c}\right),$$

$$\beta_1 = 2\lambda \sum_{n=0}^{\infty} (-1)^n (\lambda)^{-\left(\frac{2n+c+1}{c}\right)} \Gamma\left(\frac{2n+c+1}{c}\right),$$

$$\alpha_2 = 1 + 4\lambda \sum_{n=0}^{\infty} (-1)^n (n+1)(\lambda)^{-\left(\frac{2n+c+4}{c}\right)} \Gamma\left(\frac{2n+c+4}{c}\right) - 4\lambda \sum_{n=0}^{\infty} (-1)^n (\lambda)^{-\left(\frac{2n+c+2}{c}\right)} \Gamma\left(\frac{2n+c+2}{c}\right)$$

$$\beta_2 = 4\lambda \sum_{n=0}^{\infty} (-1)^n (\lambda)^{-\left(\frac{2n+c+1}{c}\right)} \Gamma\left(\frac{2n+c+1}{c}\right) - 8\lambda \sum_{n=0}^{\infty} (-1)^n (n+1)(\lambda)^{-\left(\frac{2n+c+2}{c}\right)} \Gamma\left(\frac{2n+c+2}{c}\right)$$

where

$$\int_0^{\infty} x^{\nu-1} e^{\left(-\mu x^p\right)} dx = \frac{1}{p} \mu^{-\frac{\nu}{p}} \Gamma\left(\frac{\nu}{p}\right) \text{ for } \operatorname{Re}\mu > 0, \operatorname{Re}\nu > 0 \text{ and } p > 0 \tag{6.2.5}$$

(Gradshteyn and Ryzhik, 2007, formula no. 3.478.1).

The population characteristics for the SSCWBD models can be expressed in terms of trigonometric moments α_p and β_p. From the population characteristics, it can be observed that with increasing value of parameter c, the circular variance gradually decreases and the distribution is negatively skewed and leptokurtic.

6.3 Stereographic Semicircular Half Logistic Distribution (SSCHLD) (Sreekanth et al. 2018)

A random variable X on the real line is said to have a half logistic distribution with location parameter α and scale parameter $\beta > 0$, if the probability density function and probability distribution function of X are given respectively by

$$f(x) = \frac{2}{\beta}\left[1 + \exp\left(\frac{-(x-\alpha)}{\beta}\right)\right]^{-2} \exp\left(\frac{-(x-\alpha)}{\beta}\right), \quad 0 < x < \infty, \ \beta > 0 \tag{6.3.1}$$

$$F(x) = \frac{1 - \exp\left(\frac{-(x-\alpha)}{\beta}\right)}{1 + \exp\left(\frac{-(x-\alpha)}{\beta}\right)}, \quad 0 < x < \infty \tag{6.3.2}$$

6.3.1 Probability Density Function and Cumulative Distribution Function

By applying modified inverse stereographic projection defined by a one-to-one mapping $x = v\tan\left(\frac{\theta}{2}\right)$, $v = 1$, this leads to a semicircular model on unit semicircle, called stereographic semicircular half logistic distribution.

Definition

A random variable X_{SC} on the semicircle is said to have the stereographic semicircular half logistic distribution with location parameter μ scale parameter $\sigma > 0$ denoted by **SSCHLD** (σ, μ) (Figures 6.4 through 6.6),

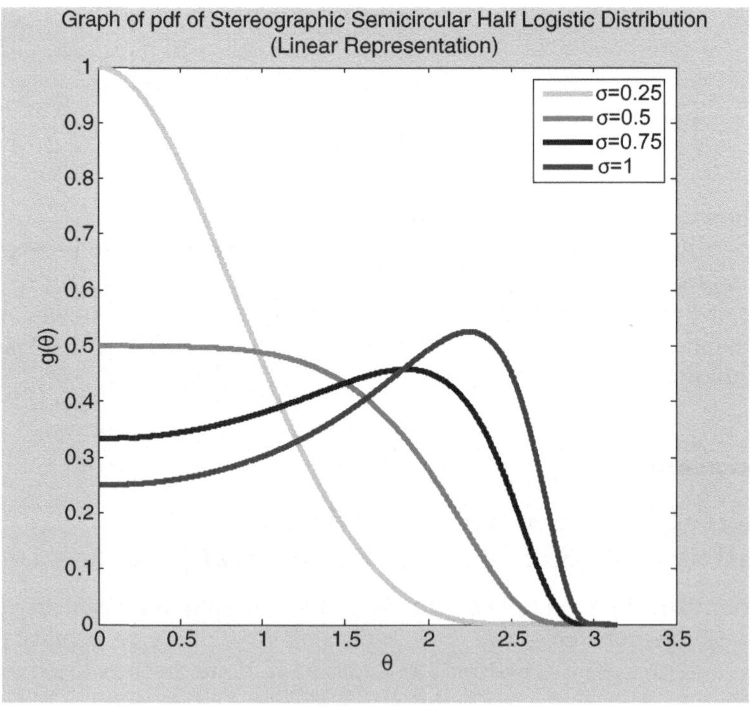

FIGURE 6.4
Graph of pdf of stereographic semicircular half logistic distribution (linear representation) for $\mu = 0$.

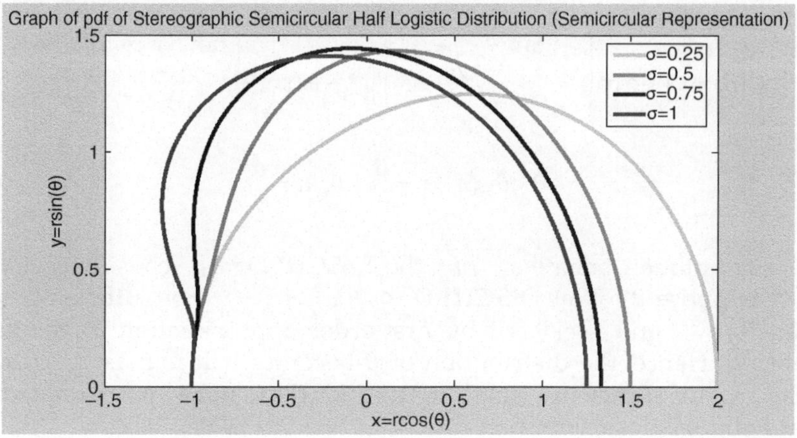

FIGURE 6.5
Graph of pdf of stereographic semicircular half logistic distribution (circular representation) for $\mu = 0$.

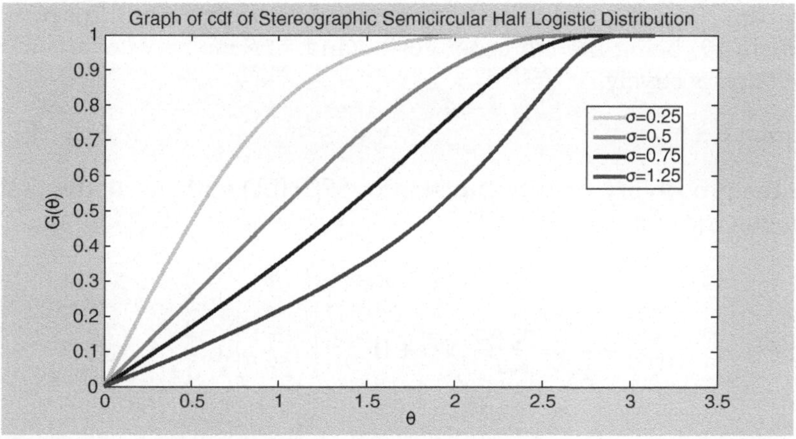

FIGURE 6.6
Graph of cdf of stereographic semicircular half logistic distribution for $\mu = 0$.

if the probability density and the cumulative distribution functions are respectively given by

$$g(\theta) = \frac{1}{\sigma}\sec^2\left(\frac{\theta}{2}\right)\left[1+\exp\left(\frac{-\left(\tan\left(\frac{\theta}{2}\right)-\mu\right)}{\sigma}\right)\right]^{-2}\exp\left(-\left(\frac{\tan\left(\frac{\theta}{2}\right)-\mu}{\sigma}\right)\right),$$

$$0<\theta<\pi,\ \sigma=\frac{\beta}{\nu}>0,\ \mu=\frac{\alpha}{\nu} \qquad (6.3.3)$$

$$G(\theta)=\left[1-\exp\left(\frac{-\left(\tan\left(\frac{\theta}{2}\right)-\mu\right)}{\sigma}\right)\right]\left[1+\exp\left(\frac{-\left(\tan\left(\frac{\theta}{2}\right)-\mu\right)}{\sigma}\right)\right]^{-1},$$

$$0<\theta<\pi,\ \sigma=\frac{\beta}{\nu}>0,\ \mu=\frac{\alpha}{\nu} \tag{6.3.4}$$

The asymptotic behavior of the SSCHLD is considered when $\sigma \to 0$. Suppose θ follows **SSCHLD** (σ,μ). Let $y=\frac{\theta}{\sigma}$. For sufficiently small σ, $\tan\left(\frac{\sigma y}{2}\right)\approx\frac{\sigma y}{2}$ and $\sec\left(\frac{\sigma y}{2}\right)\approx 1$ by first-order approximation of the Taylor expansion. Hence, the distribution of θ becomes half logistic distribution (linear). So, for sufficiently small σ, the SSCHLD can be approximated by a linear half logistic distribution.

6.3.2 Characteristic Function and Trigonometric Moments

The trigonometric moments of the distribution are given by $\{\varphi_p : p=\pm1,\pm2,\pm3,\ldots\}$, where $\varphi_p=\alpha_p+i\beta_p$, with $\alpha_p=E(\cos p\theta)$ and $\beta_p=E(\sin p\theta)$ being the pth order cosine and sine moments of the random angle θ, respectively.

Theorem 6.3.1

With the probability density function of SSCHLD with $\mu=0$, the first two trigonometric moments are

$$\alpha_1=1-\frac{2}{\sigma\sqrt{\pi}}\sum_{n=0}^{\infty}(-1)^n(n+1)G_{13}^{31}\left(\left(\frac{n+1}{2\sigma}\right)^2\middle|\begin{matrix}-\frac{1}{2}\\-\frac{1}{2},0,\frac{1}{2}\end{matrix}\right),$$

$$\beta_1=\frac{2}{\sigma\sqrt{\pi}}\sum_{n=0}^{\infty}(-1)^n(n+1)G_{13}^{31}\left(\left(\frac{n+1}{2\sigma}\right)^2\middle|\begin{matrix}0\\0,0,\frac{1}{2}\end{matrix}\right),$$

$$\alpha_2=1+\frac{2}{\sigma\sqrt{\pi}}$$

$$\sum_{n=0}^{\infty}(-1)^n(n+1)\left[G_{13}^{31}\left(\left(\frac{n+1}{2\sigma}\right)^2\middle|\begin{matrix}-\frac{3}{2}\\-\frac{1}{2},0,\frac{1}{2}\end{matrix}\right)-G_{13}^{31}\left(\left(\frac{n+1}{2\sigma}\right)^2\middle|\begin{matrix}-\frac{1}{2}\\-\frac{1}{2},0,\frac{1}{2}\end{matrix}\right)\right],$$

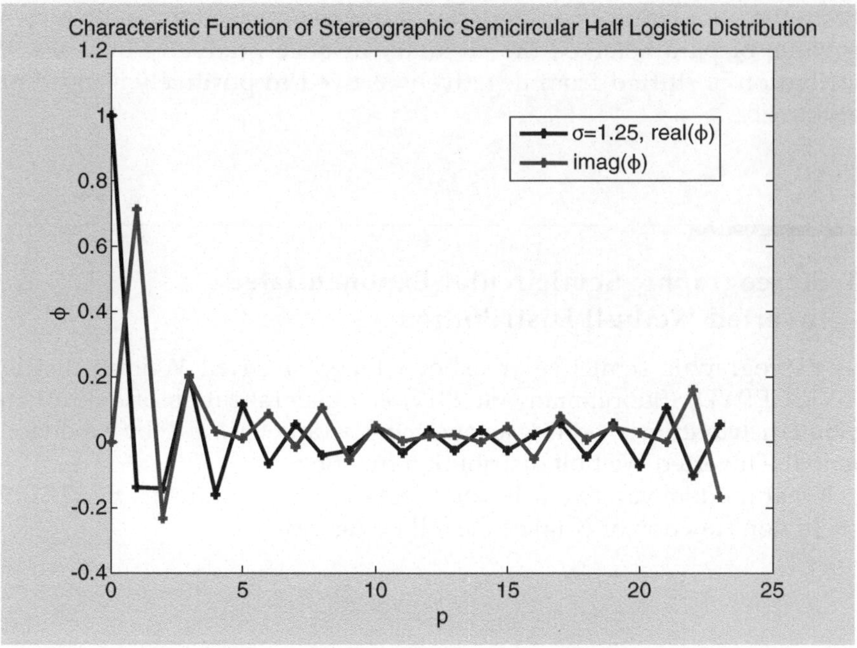

FIGURE 6.7
Graph of characteristic function of stereographic semicircular half logistic distribution for $\sigma=1.25$.

$$\beta_2 = \frac{4}{\sigma\sqrt{\pi}}\sum_{n=0}^{\infty}(-1)^n(n+1)\left[G_{13}^{31}\left(\left(\frac{n+1}{2\sigma}\right)^2\middle|\begin{matrix}0\\0,0,\frac{1}{2}\end{matrix}\right)-4G_{13}^{31}\left(\left(\frac{n+1}{2\sigma}\right)^2\middle|\begin{matrix}-1\\0,0,\frac{1}{2}\end{matrix}\right)\right].$$

The first two trigonometric moments are sufficient for calculating population characteristics (Figure 6.7)

where $\int_0^\infty x^{2\nu-1}\left(u^2+x^2\right)^{Q-1}e^{-\mu x}dx=\frac{u^{2\nu+2Q-2}}{2\sqrt{\pi}\Gamma(1-Q)}G_{13}^{31}\left(\frac{\mu^2u^2}{4}\middle|\begin{matrix}1-\nu\\1-Q-\nu,0,\frac{1}{2}\end{matrix}\right)$

for $|\arg u\pi|<\frac{\pi}{2}$, $\operatorname{Re}\mu>0$ and $\operatorname{Re}\nu>0$ and $G_{13}^{31}\left(\frac{\mu^2u^2}{4}\middle|\begin{matrix}1-\nu\\1-Q-\nu,0,\frac{1}{2}\end{matrix}\right)$ is called as

Meijer's G-function (Gradshteyn and Ryzhik 2007, formula no. 3.389.2).

The characteristics for the Stereographic Semicircular Half Logistic distribution are also worked out using their respective trigonometric moments.

From the population characteristics, it can be observed that with increasing value of parameter σ, the circular variance gradually increase, the distribution is shifted from negatively skewed to positively skewed and platykurtic.

6.4 Stereographic Semicircular Exponentiated Inverted Weibull Distribution

The stereographic semicircular exponentiated inverted Weibull distribution (SSCEIWD) (Subrahmanyam 2017) is a generalization of the inverted Weibull distribution by adding a new shape parameter $\lambda \in R^{+}$ by exponentiation to the inverted Weibull distribution function.

A linear random variable X is said to follow a two-parameter EIWD, if the distribution function of X takes the following form

$$F(x) = \left(e^{-x^{-c}}\right)^{\lambda} \tag{6.4.1}$$

where c and λ both are shape parameters and $0 < x < \infty$ and $c > 0, \lambda > 0$.

Hence, the probability density function of SSCEIWD is

$$f(x) = \lambda\, c\, x^{-(c+1)}\left(e^{-x^{-c}}\right)^{\lambda} \tag{6.4.2}$$

where $0 < x < \infty$ and $c > 0, \lambda > 0$.

Here, if $\lambda = 1$, this SSCEIWD becomes the standard inverted Weibull distribution and if $c = 1$, SSCEIWD represents standard inverted exponential distribution.

6.4.1 Probability Density Function and Cumulative Distributive Function

By applying the inverse stereographic projection on the exponentiated inverted Weibull distribution (Figures 6.8 through 6.10), the one-to-one mapping of $T(\theta) = x = \tan\left(\frac{\theta}{2}\right)$; where $x \in (0, \infty)$, $\theta \in (0, \pi)$ yields the stereographic semicircular exponentiated inverted Weibull distribution (SSCEIWD) with probability density function $g(\theta)$ can be obtained as below:

$$g(\theta) = \tfrac{1}{2}\sec^{2}\left(\tfrac{\theta}{2}\right)\lambda\, c\left(\tan\left(\tfrac{\theta}{2}\right)\right)^{-(c+1)}\left(e^{-\left(\tan\left(\frac{\theta}{2}\right)\right)^{-c}}\right)^{\lambda} \tag{6.4.3}$$

where $\theta \in (0, \pi)$, $c > 0, \lambda > 0$.

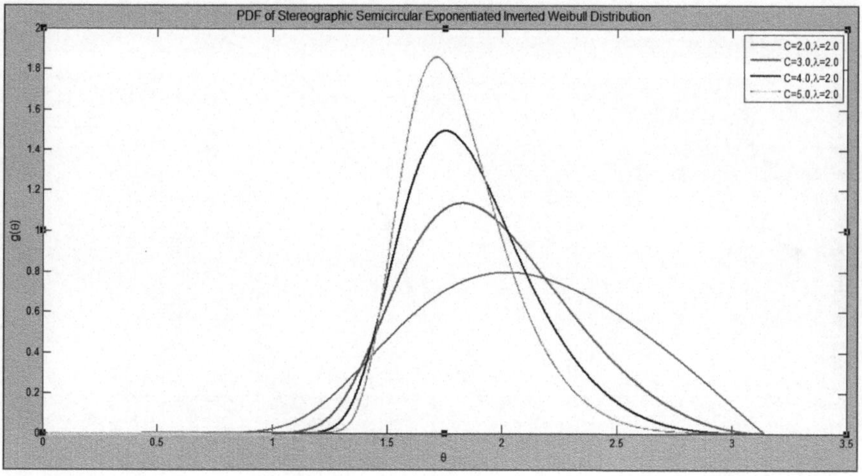

FIGURE 6.8
Graph of pdf of SSCEIWD (linear representation for different values of c and $\lambda = 2$).

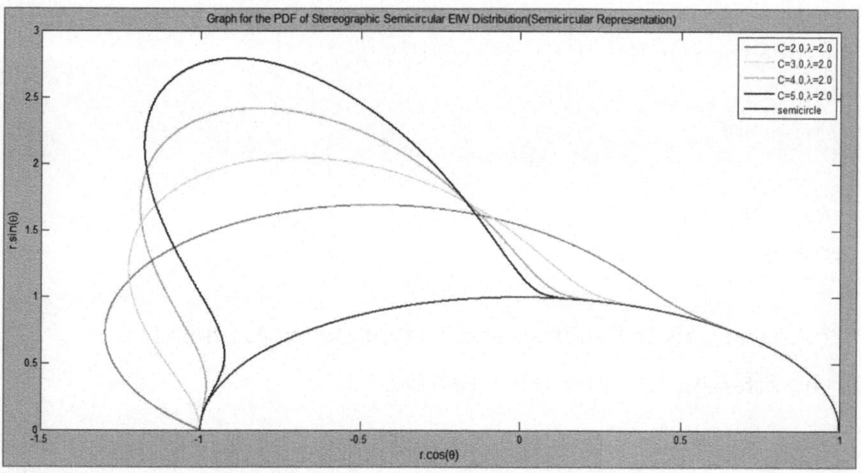

FIGURE 6.9
Graph of pdf of SSCEIWD (semicircular representation for different values of c and $\lambda = 2$).

The cdf, $G(\theta)$ for the SSCEIWD can be derived as follows:

$$G(\theta) = \int_0^\theta g(\theta) d\theta$$

$$= \int_0^\theta \frac{1}{2}\sec^2\left(\frac{\theta}{2}\right)\lambda c\left(\tan\left(\frac{\theta}{2}\right)\right)^{-(c+1)}\left(e^{-\left(\tan\left(\frac{\theta}{2}\right)\right)^{-c}}\right)^{\lambda} d\theta \tag{6.4.4}$$

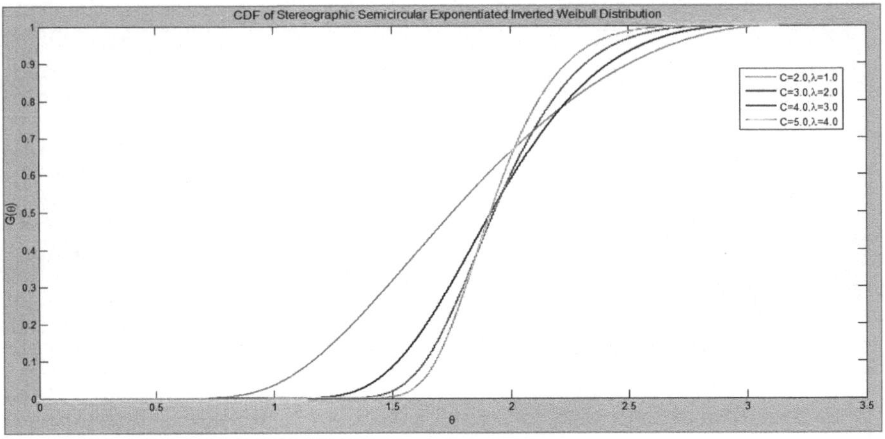

FIGURE 6.10
Graph of cdf of SSCEIWD for various values of c and λ.

Evaluating the equation (6.4.4), we get $G(\theta)$ as

$$G(\theta) = \left(e^{-\left(\tan\left(\frac{\theta}{2}\right)\right)^{-c}} \right)^{\lambda} \tag{6.4.5}$$

where $\theta \in (0, \pi), c > 0, \lambda > 0$.

6.4.2 Characteristic Function and Trigonometric Moments

The characteristic function of SSCEIWD is

$$\varphi_{sc}(p) = \int_0^{\pi} e^{ip\theta}\, \tfrac{1}{2} \sec^2\left(\tfrac{\theta}{2}\right) \lambda\, c \left(\tan\left(\tfrac{\theta}{2}\right)\right)^{-(c+1)} \left(e^{-\left(\tan\left(\frac{\theta}{2}\right)\right)^{-c}} \right)^{\lambda} d\theta \tag{6.4.6}$$

Let $y = \left(\tan\left(\frac{\theta}{2}\right)\right)^{-c}$ then

$$\varphi_{sc}(p) = \int_0^{\infty} e^{ip\left(2\tan^{-1}\left(y^{\frac{-1}{c}}\right)\right)} \lambda\, e^{-y\lambda}\, dy$$

again considering $y\lambda = k$ and making necessary transformations to the above equation, it becomes

$$\varphi_{sc}(p) = \int_0^{\infty} e^{ip\left(2\tan^{-1}\left(\left(\frac{k}{\lambda}\right)^{-1/c}\right)\right)} e^{-k} dk \tag{6.4.7}$$

Equation (6.4.7) is evaluated, and trigonometric moments for SSCEIWD are obtained. The population characteristics of SSCEIWD are computed using these trigonometric moments. The following is the graph for the characteristic function of the SSCEIWD showing the real and imaginary parts α_p and β_p separately for different values of c and λ (Figure 6.11).

From the characteristic function of the stereographic semicircular model obtained above, the first and second trigonometric moments for SSCEIWD distribution are derived as follows:

$$\varphi_{sc}(p) = \int_0^{\pi} e^{ip\theta} g(\theta) d\theta,\ p \in \mathbb{Z} \text{ this can also be written as}$$

$$\varphi_{sc}(p) = \int_0^{\pi} \cos p\theta\ g(\theta) d\theta + i\int_0^{\pi} \sin p\theta\ g(\theta) d\theta,\ p \in \mathbb{Z} \tag{6.4.8}$$

Considering

$$\alpha_p = \int_0^{\pi} \cos p\theta\ g(\theta) d\theta \qquad \beta_p = \int_0^{\pi} \sin p\theta\ g(\theta) d\theta$$

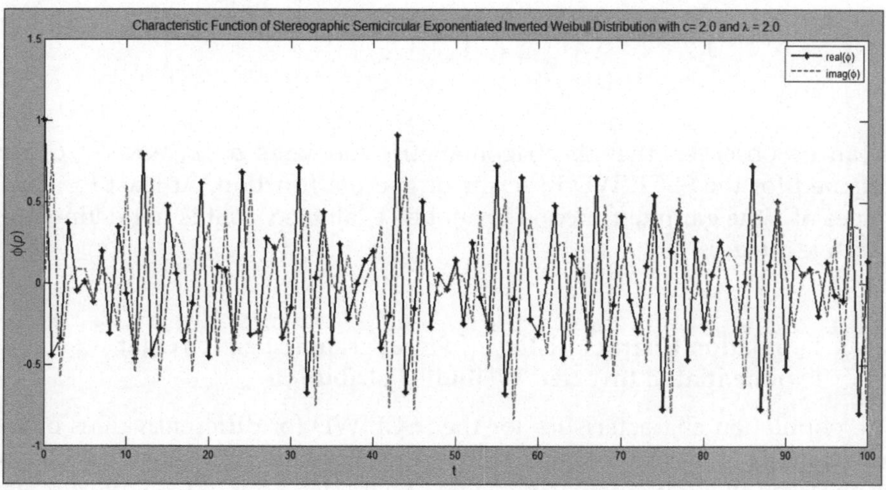

FIGURE 6.11
Graph of the characteristic function of SSCEIWD at $c = 2$ and $\lambda = 2$.

By putting $p=1$ and $p=2$, in the above equations first and second cosine and sine moments are derived for the SSCEIWD from the characteristic function.

$$\phi_{sc}(p)=\int_0^{\pi} e^{ip\theta}\,\tfrac{1}{2}\sec^2\left(\tfrac{\theta}{2}\right)\lambda c\left(\tan\left(\tfrac{\theta}{2}\right)\right)^{-(c+1)}\left(e^{-\left(\tan\left(\frac{\theta}{2}\right)\right)^{-c}}\right)^{\lambda} d\theta \tag{6.4.9}$$

Theorem 6.4.1

With the probability density function of SSCEIWD with $\mu=0$, the first two trigonometric moments are

$$\alpha_1=\int_0^{\pi}\frac{1}{2}\cos\theta\,\sec^2\left(\frac{\theta}{2}\right)\lambda c\left(\tan\left(\frac{\theta}{2}\right)\right)^{-(c+1)}\left(e^{-\left(\tan\left(\frac{\theta}{2}\right)\right)^{-c}}\right)^{\lambda} d\theta. \tag{6.4.10}$$

$$\beta_1=\int_0^{\pi}\frac{1}{2}\sin\theta\,\sec^2\left(\frac{\theta}{2}\right)\lambda c\left(\tan\left(\frac{\theta}{2}\right)\right)^{-(c+1)}\left(e^{-\left(\tan\left(\frac{\theta}{2}\right)\right)^{-c}}\right)^{\lambda} d\theta \tag{6.4.11}$$

$$\alpha_2=\int_0^{\pi}\frac{1}{2}(\cos 2\theta)\sec^2\left(\frac{\theta}{2}\right)\lambda c\left(\tan\left(\frac{\theta}{2}\right)\right)^{-(c+1)}\left(e^{-\left(\tan\left(\frac{\theta}{2}\right)\right)^{-c}}\right)^{\lambda} d\theta \tag{6.4.12}$$

$$\beta_2=\int_0^{\infty}\frac{1}{2}\sin 2\theta\,\sec^2\left(\frac{\theta}{2}\right)\lambda c\left(\tan\left(\frac{\theta}{2}\right)\right)^{-(c+1)}\left(e^{-\left(\tan\left(\frac{\theta}{2}\right)\right)^{-c}}\right)^{\lambda} d\theta \tag{6.4.13}$$

It can be observed that the trigonometric moments α_1, β_1 and α_2, β_2 are obtained for the SSCEIWD in terms of gamma function. At least for some values of c the gamma function cannot be evaluated, as it is not defined for negative integers.

6.4.3 Population Characteristics of Stereographic Semicircular Exponentiated Inverted Weibull Distribution

The population characteristics for the SSCEIWD for different values of the parameters c and λ are computed. From the population characteristics, it can be observed that with increasing value of parameter c and $\lambda=3$, the circular variance gradually decreases and the distribution is shifted from positively skewed to negatively skewed and platykurtic.

6.5 Arc Offset Beta Model

An arc model directly by applying offsetting on a linear bivariate model is obtained, irrespective of imposing restriction on circular random variable. It is a natural occurrence. The arc offset beta (AOB) distribution (Radhika 2014) is derived from the bivariate beta distribution (Balakrishnan and Chin-Diew 2008, p. 374).

6.5.1 Probability Density Function and Cumulative Distribution Function of Offset Beta Model

$$f(x,y)=\frac{\Gamma(a+b+c)}{\Gamma a\,\Gamma b\,\Gamma c}\cdot x^{a-1}\;y^{b-1}(1-x-y)^{c-1},\qquad x+y\le 1\,,\,0\le x,\,y\le 1\text{and }a,b,c>0 \tag{6.5.1}$$

By applying offsetting, x and y are transformed as $x=r\cos\theta,\,y=r\sin\theta$.

When $0<x,\,y<1$, we have $0<r\cos\theta,r\sin\theta<1$ which results in $0\le\theta<\frac{\pi}{2}$. Hence, by applying offsetting,

$$g(\theta)=\int_0^\infty f(r\cos\theta,r\sin\theta)\,r\,dr$$

$$=\int_0^\infty\frac{\Gamma(a+b+c)}{\Gamma a\,\Gamma b\,\Gamma c}(r\cos\theta)^{a-1}(r\sin\theta)^{b-1}\left(1-r(\cos\theta+\sin\theta)^{c-1}\right)r\,dr$$

$$=\frac{\Gamma(a+b+c)}{\Gamma a\,\Gamma b\,\Gamma c}\cos^{a-1}\theta\,\sin^{b-1}\theta\int_0^\infty r^{a+b-1}\left(1-(r\cos\theta+\sin\theta)^{c-1}\right)dr$$

$$=\frac{\Gamma(a+b+c)}{\Gamma a\,\Gamma b\,\Gamma c}\,\frac{\cos\theta^{a-1}\,\sin\theta^{b-1}}{(\cos\theta+\sin\theta)^{a+b}}\,\beta(a+b,c)$$

The pdf of the AOB model is

$$g(\theta)=\frac{\Gamma(a+b)}{\Gamma a\;\Gamma b}\,\frac{\cos^{(a-1)}\theta\,\sin^{(b-1)}\theta}{(\cos\theta+\sin\theta)^{a+b}}$$

$$=\frac{1}{\beta(a,b)}\frac{\cos^{(a-1)}\theta\,\sin^{(b-1)}\theta}{(\cos\theta+\sin\theta)^{a+b}},\quad 0\le\theta\le\frac{\pi}{2}.$$

The cdf of the AOB model is

$$G(\theta) = \int_0^{\theta} g(\alpha)d\alpha$$

$$= \frac{1}{\beta(a,b)} \int_0^{\theta} \frac{\cos^{(a-1)}\alpha \, \sin^{(b-1)}\alpha}{(\cos\alpha + \sin\alpha)^{a+b}} d\alpha, \, 0 \le \alpha \le \frac{\pi}{2}$$

$$= 1 - \frac{\beta_p\,(a,b)}{\beta\,(a,b)}$$

where $p = \dfrac{1}{1+\tan\theta}$

Definition 6.5.1

A random variable X_{AO} on the unit circle is said to have offset beta distribution with parameters a and b denoted by $\text{AOB}(a,b)$, if the probability density and cumulative distribution functions are respectively given by

$$g(\theta) = \frac{1}{\beta(a,b)} \frac{\cos^{(a-1)}\theta \, \sin^{(b-1)}\theta}{(\cos\theta + \sin\theta)^{a+b}}, \quad 0 \le \theta \le \frac{\pi}{2} \tag{6.5.2}$$

and

$$G(\theta) = 1 - \frac{\beta_p\,(a,b)}{\beta\,(a,b)} \quad \text{where } p = \frac{1}{1+\tan\theta} \tag{6.5.3}$$

where $\beta_p\,(a,b)$ is incomplete beta function.

The graphs of the pdf (Figure 6.12) and cdf (Figure 6.13) of the AOB model for various values of a and b are plotted here.

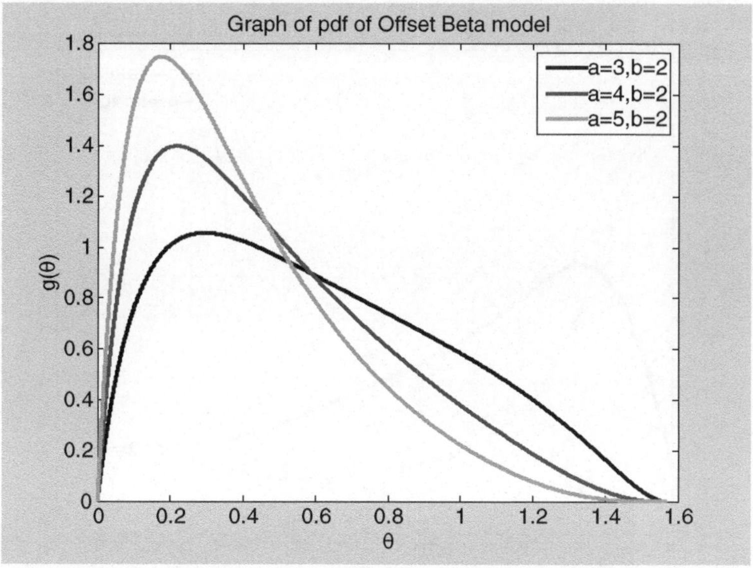

FIGURE 6.12
Graph of pdf of the arc offset beta model.

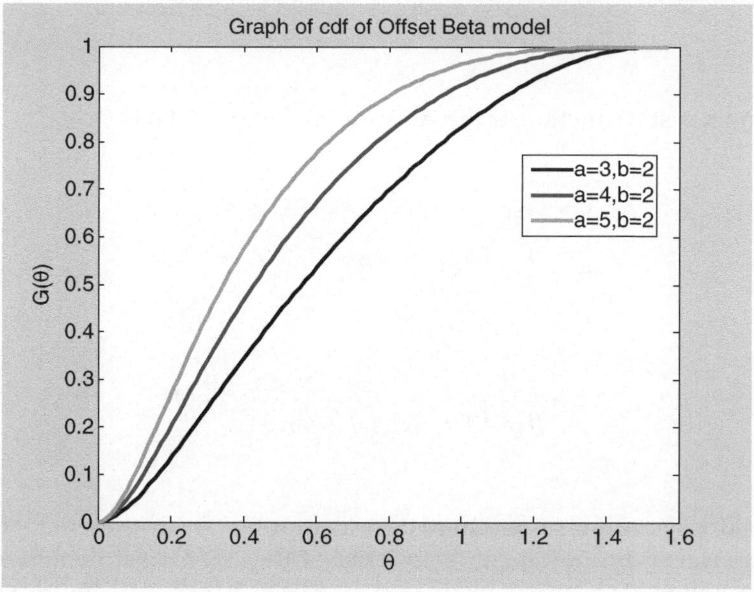

FIGURE 6.13
Graph of cdf of the arc offset beta model.

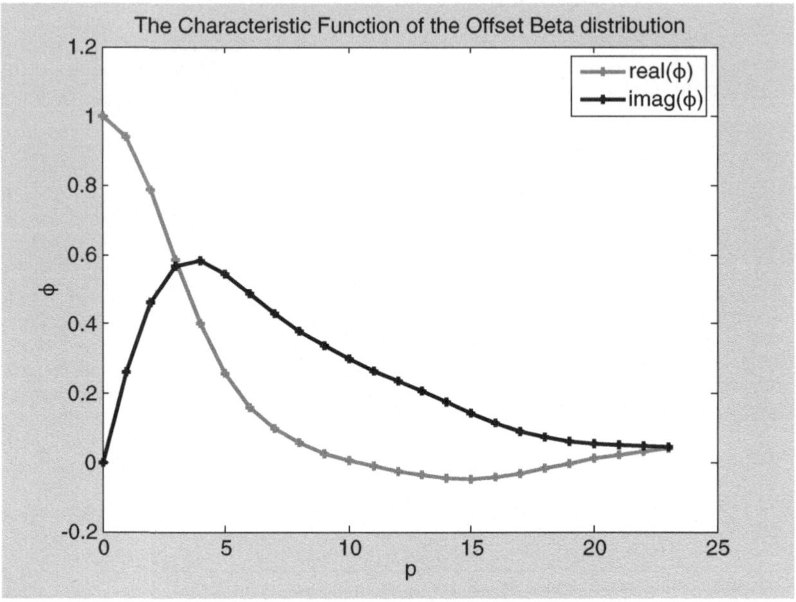

FIGURE 6.14
Graph of the characteristic function of the arc offset beta distribution.

the characteristic function of the AOB model (Figure 6.14) is

$$\varphi_p = \int_0^{\frac{\pi}{2}} e^{ip\theta} g(\theta)\ d\theta \qquad p \in \mathbb{Z}$$

$$= \frac{1}{\beta(a,b)} \int_0^{\frac{\pi}{2}} \frac{e^{ip\theta} \cos^{a-1}\theta \ \sin^{b-1}\theta}{(\cos\theta + \sin\theta)^{a+b}}\ d\theta$$

Required trigonometric moments are computed from the characteristic function. The population characteristics of the Arc Offset Beta model for various values of a and b evaluated in MATLAB. From the population characteristics, it can be observed that with increasing value of parameter a and fixed value of b, the circular variance gradually decreases and the distribution is negatively skewed and shifted from platykurtic to leptokurtic.

6.6 Other Extemporaneous Semicircular/Arc Models

6.6.1 Stereographic Semicircular Exponential Distribution (SSCEXPD) (Phani 2013)

1. Probability Density Function

$$g(\theta)=\frac{\lambda}{2}\sec^2\left(\frac{\theta}{2}\right)\exp\left(-\lambda\left(\tan\left(\frac{\theta}{2}\right)-\mu\right)\right), \text{ for } 0\le\theta<\pi,\ \mu=\frac{\gamma}{v} \text{ and}$$

$\lambda=\sigma v>0$

2. Cumulative Distribution Function

$$G(\theta)=1-e^{-\lambda\left(\tan\left(\frac{\theta}{2}\right)-\mu\right)}$$

3. Characteristic Function and Trigonometric Moments

Without loss of generality, here we assume that $\mu=0$, in the pdf of the SSCEXPD model. The trigonometric moments of the distribution are given by $\{\varphi_p: p=\pm1,\pm2,\pm3,\ldots\}$, where $\varphi_p=\alpha_p+\beta_p$, with $\alpha_p=E(\cos p\theta)$ and $\beta_p=E(\sin p\theta)$ being the *p*th order cosine and sine moments of the random angle θ, respectively.

Under the pdf of the SSCEXPD model with $\mu=0$, the first two $\alpha_p=E(\cos p\theta)$ and $\beta_p=E(\sin p\theta)$, $p=1,2$ are given as follows:

$$\alpha_1=1-\frac{\lambda}{\sqrt{\pi}}G_{13}^{31}\left(\frac{\lambda^2}{4}\middle|\begin{matrix}-\frac{1}{2}\\ -\frac{1}{2},0,\frac{1}{2}\end{matrix}\right)$$

$$\beta_1=\frac{\lambda}{\sqrt{\pi}}G_{13}^{31}\left(\frac{\lambda^2}{4}\middle|\begin{matrix}0\\ 0,0,\frac{1}{2}\end{matrix}\right)$$

$$\alpha_2=1+\frac{2\lambda}{\sqrt{\pi}}G_{13}^{31}\left(\frac{\lambda^2}{4}\middle|\begin{matrix}-\frac{3}{2}\\ -\frac{1}{2},0,\frac{1}{2}\end{matrix}\right)-\frac{2\lambda}{\sqrt{\pi}}G_{13}^{31}\left(\frac{\lambda^2}{4}\middle|\begin{matrix}-\frac{1}{2}\\ -\frac{1}{2},0,\frac{1}{2}\end{matrix}\right)$$

$$\beta_2 = \frac{2\lambda}{\sqrt{\pi}} G_{13}^{31}\left(\frac{\lambda^2}{4}\middle|\begin{matrix}0\\0,0,\frac{1}{2}\end{matrix}\right) - \frac{4\lambda}{\sqrt{\pi}} G_{13}^{31}\left(\frac{\lambda^2}{4}\middle|\begin{matrix}-1\\0,0,\frac{1}{2}\end{matrix}\right)$$

where $\int_0^\infty x^{2v-1}\left(u^2+x^2\right)^{Q-1} e^{-\mu x}dx = \frac{u^{2v+2Q-2}}{2\sqrt{\pi}\Gamma(1-Q)} G_{13}^{31}\left(\frac{\mu^2u^2}{4}\middle|\begin{matrix}1-v\\1-Q-v,0,\frac{1}{2}\end{matrix}\right)$

for $|\arg u\pi| < \frac{\pi}{2}$, $\operatorname{Re}\mu > 0$ and $\operatorname{Re}v > 0$ and $G_{13}^{31}\left(\frac{\mu^2u^2}{4}\middle|\begin{matrix}1-v\\1-Q-v,0,\frac{1}{2}\end{matrix}\right)$ is called as Meijer's G-function (Gradshteyn and Ryzhik 2007, formula no. 3.389.2).

6.6.2 Stereographic Semicircular Gamma Distribution (SSCGD) Model (Radhika 2014)

1. Probability Density Function

$$g(\theta) = \frac{1}{2\sigma^c\Gamma(c)} \sec^2\left(\frac{\theta}{2}\right)\left(\tan\left(\frac{\theta}{2}\right)-\mu\right)^{c-1} \exp\left(-\frac{1}{\sigma}\left(\tan\left(\frac{\theta}{2}\right)-\mu\right)\right),$$

where $0 \le \theta < \pi, c > 0, \sigma = \frac{\lambda}{v} > 0$ and $\mu = \frac{\alpha}{v}$

2. Cumulative Distribution Function

$$G(\theta) = \frac{\Gamma_t(c)}{\Gamma(c)} \text{ where } t = \frac{1}{\sigma}\tan\left(\frac{\theta}{2}\right), \theta \in [0,\pi)$$

3. Characteristic Function and Trigonometric Moments

$$\varphi_{X_{SC}}(p) = \int_0^\pi e^{ip\theta} g(\theta)d\theta$$

$$= \frac{1}{2\sigma^c\Gamma(c)}\int_0^\pi e^{ip\theta}\sec^2\left(\frac{\theta}{2}\right)\left(\tan\left(\frac{\theta}{2}\right)-\mu\right)^{c-1} \exp\left(-\frac{1}{\sigma}\left(\tan\left(\frac{\theta}{2}\right)-\mu\right)\right)d\theta$$

$$= \frac{1}{2\sigma^c\Gamma(c)}\int_0^\pi e^{ip\theta}\sec^2\left(\frac{\theta}{2}\right)\tan\left(\frac{\theta}{2}\right)^{c-1} e^{-\frac{1}{\sigma}\tan\left(\frac{\theta}{2}\right)}d\theta \text{ for } \mu=0$$

The trigonometric moments of the distribution are given by $\{\varphi_p : p = \pm 1, \pm 2, \pm 3, \ldots\}$, where $\varphi_p = \alpha_p + \beta_p$, with $\alpha_p = E(\cos p\theta)$ and $\beta_p = E(\sin p\theta)$ being the pth order cosine and sine moments of the random angle θ, respectively.

The first two $\alpha_{\mathrm{p}} = E(\cos \mathrm{p}\theta)$ and $\beta_{\mathrm{p}} = E(\sin \mathrm{p}\theta)$, p=1,2 of the SSCGD with $\mu = 0$, are given as follows:

$$\alpha_1 = 1 - \frac{1}{\sigma^c \sqrt{\pi}\,\Gamma(c)} G_{13}^{31}\left(\frac{1}{4\sigma^2} \middle| \begin{matrix} -\frac{c}{2} \\ -\frac{c}{2}, 0, \frac{1}{2} \end{matrix} \right)$$

$$\beta_1 = \frac{1}{\sigma^c \sqrt{\pi}\,\Gamma(c)} G_{13}^{31}\left(\frac{1}{4\sigma^2} \middle| \begin{matrix} \frac{1-c}{2} \\ \frac{1-c}{2}, 0, \frac{1}{2} \end{matrix} \right)$$

$$\alpha_2 = 1 + \frac{4}{\sigma^c \sqrt{\pi}\,\Gamma(c)} G_{13}^{31}\left(\frac{1}{4\sigma^2} \middle| \begin{matrix} -\frac{c}{2} - 1 \\ -\frac{c}{2}, 0, \frac{1}{2} \end{matrix} \right) - \frac{4}{\sigma^c \sqrt{\pi}\,\Gamma(\lambda)} G_{13}^{31}\left(\frac{1}{4\sigma^2} \middle| \begin{matrix} -\frac{c}{2} \\ -\frac{c}{2}, 0, \frac{1}{2} \end{matrix} \right)$$

$$\beta_2 = \frac{2}{\sigma^c \sqrt{\pi}\,\Gamma(\lambda)} G_{13}^{31}\left(\frac{1}{4\sigma^2} \middle| \begin{matrix} -\frac{c}{2} \\ -\frac{c}{2}, 0, \frac{1}{2} \end{matrix} \right) - \frac{4}{\sigma^c \sqrt{\pi}\,\Gamma(\lambda)} G_{13}^{31}\left(\frac{1}{4\sigma^2} \middle| \begin{matrix} -\frac{(c+1)}{2} \\ -\frac{(1-c)}{2}, 0, \frac{1}{2} \end{matrix} \right)$$

6.6.3 Stereographic Semicircular New Weibull–Pareto Distribution (SSCNWPD) (Srinivasa Subrahmanyam et al. 2017)

1. Probability Density Function

$$g(\theta) = \frac{1}{2} \operatorname{Sec}^2\left(\tfrac{\theta}{2}\right) \frac{c\,\delta}{\lambda} \left(\frac{\tan\left(\frac{\theta}{2}\right)}{\lambda} \right)^{c-1} e^{-\delta \left(\frac{\tan\left(\frac{\theta}{2}\right)}{\lambda} \right)^{c}}$$

where $\theta \in (0, \pi), c > 0, \lambda > 0$, and $\delta > 0$.

2. Cumulative Distribution Function

where $\theta \in (0, \pi), c > 0, \lambda > 0$, and $\delta > 0$

3. Characteristic Function and Trigonometric Moments

$$\phi_{sc}(p)=\int_0^{\pi} e^{ip\theta}\frac{1}{2}\mathrm{Sec}^2\left(\frac{\theta}{2}\right)\frac{c\,\delta}{\lambda}\left(\frac{\tan\left(\frac{\theta}{2}\right)}{\lambda}\right)^{c-1} e^{-\delta\left(\frac{\tan\left(\frac{\theta}{2}\right)}{\lambda}\right)^{c}}\,d\theta$$

The trigonometric moments of the distribution are given by $\{\varphi_p : p=\pm1,\pm2,\pm3,\ldots\}$, where $\varphi_p=\alpha_p+\beta_p$, with $\alpha_p=E(\cos p\theta)$ and $\beta_p=E(\sin p\theta)$ being the pth order cosine and sine moments of the random angle θ, respectively.

The first two $\alpha_p=E(\cos p\theta)$ and $\beta_p=E(\sin p\theta)$, p=1,2 of the SSCNWPD with $\mu=0$ are given as follows:

$$\alpha_1=1-2\sum_{n=0}^{\infty}(-1)^n\left(\frac{\lambda}{\delta^{\frac{1}{c}}}\right)^{2n+2}\Gamma\left(1+\frac{(2n+2)}{c}\right)$$

$$\beta_1=2\sum_{n=0}^{\infty}(-1)^n\left(\frac{\lambda}{\delta^{\frac{1}{c}}}\right)^{2n+1}\Gamma\left(1+\frac{(2n+1)}{c}\right)$$

$$\alpha_2=1-8\sum_{n=0}^{\infty}(-1)^n(n+1)\left(\frac{\lambda}{\delta^{\frac{1}{c}}}\right)^{2n+2}\Gamma\left(1+\frac{(2n+2)}{c}\right)$$

$$\beta_2=4\sum_{n=0}^{\infty}(-1)^n\left(\frac{\lambda}{\delta^{\frac{1}{c}}}\right)^{2n+1}\Gamma\left(1+\frac{2n+1}{c}\right)$$

$$-8\sum_{n=0}^{\infty}(-1)^n(n+1)\left(\frac{\lambda}{\delta^{\frac{1}{c}}}\right)^{2n+3}\Gamma\left(1+\frac{(2n+3)}{c}\right)$$

6.6.4 Arc Offset Exponential (AOEXP) Type Model (Radhika 2014)

1. Probability Density Function

$$g(\theta)=\frac{1+2\rho}{(\cos\theta+\sin\theta)^2}-\frac{2\rho}{(2\cos\theta+\sin\theta)^2}-\frac{2\rho}{(\cos\theta+2\sin\theta)^2}\text{, and}$$

2. Cumulative Distribution Function

$$G(\theta)=1+\frac{2\rho}{\tan\theta+2}+\frac{\rho}{2\tan\theta+1}-\frac{1+2\rho}{1+\tan\theta}$$

where $-1<\rho<1$ and $\theta\in\left[0,\frac{\pi}{2}\right]$

3. Characteristic Function and Trigonometric Moments

For $p\in\mathbb{Z}$, the characteristic function of the AOEXP model is

$$\varphi_p=\int_0^{\frac{\pi}{2}} e^{ip\theta}\,g(\theta)\;d\theta$$

$$=\int_0^{\frac{\pi}{2}} e^{ip\theta}\left(\frac{1+2\rho}{(\cos\theta+\sin\theta)^2}-\frac{2\rho}{(2\cos\theta+\sin\theta)^2}-\frac{2\rho}{(\cos\theta+2\sin\theta)^2}\right)d\theta$$

where $-1<\rho<1$ and $\theta\in\left[0,\frac{\pi}{2}\right]$.

References

Balakrishnan, N. and Chin-Diew, L.A.I. 2008. *Continuous Bivariate Distributions*, Springer, New York.

Fisher, N.I. 1993. *Statistical Analysis of Circular Data*, Cambridge University Press, Cambridge.

Gradshteyn, I.S. and Ryzhik, I.M. 2007. *Table of Integrals, Series and Products*, 7th ed., Academic Press, London, UK.

Jammalamadaka, S.R. and Sen Gupta, A. 2001. *Topics in Circular Statistics*, World Scientific Press, Singapore.

Minh, D.P. and Farnum, N.R. 2003. Using Bilinear Transformation to Induce Probability Distributions, *Communications in Statistics-Theory and Methods*, 32(1), 1–9.

Phani, Y. 2013. On Stereographic Circular and Semicircular Models, PhD Dissertation, Acharya Nagarjuna University, Guntur, India.

Phani, Y., Girija, S.V.S. and Dattatreya Rao, A.V. 2013. On Construction of Stereographic Semicircular Models, *Journal of Applied Probability and Statistics,* 8(1), 75–90.

Radhika, A.J.V. 2014. Mathematical Tools in the Construction of New Circular Models, PhD dissertation, Acharya Nagarjuna University, Guntur, India.

Smith, N.M. 1988. Reconstruction of the Tertiary Drainage Systems of the Inverellregion. Unpublished B.Sc. (Hons.) thesis, Department of Geography, University of Sidney, Australia.

Sreekanth, Y., Phani, Y., Girija, S.V.S. and Dattatreya Rao, A.V. 2018. Stereographic *l*: Axial Half Logistic Distribution, *International Journal of Applied Engineering Research (IJAER),* 13(12), 10627–10634.

Srinivasa Subrahmanyam, P., Dattatreya Rao, A.V. and Girija, S.V.S. 2017. On Stereographic Semicircular New Weibull Pareto Model, *IJIRST–International Journal for Innovative Research in Science & Technology,* 3(11), 267–276.

Subrahmanyam, P.S. 2017. On Construction of Angular Models with Applications to Control Charts, PhD dissertation, Acharya Nagarjuna University, Guntur, India.

7

Asymmetric l-*Axial Models*

7.1 Introduction

Marshall and Olkin (1997) proposed an interesting method of adding a new parameter to the exponential and the Weibull family of distributions. The resulting distributions are called the Marshall–Olkin extended distributions. These distributions include the original distributions as special cases and are more flexible and represent a wide range of behavior than the original distributions, particularly, in the context of life testing models.

The idea of Marshall–Olkin is suitably adopted for circular data. Using this, a new class of asymmetric Marshall–Olkin stereographic and wrapped Marshall–Olkin, circular models are derived. The construction of these types of models is demonstrated on logistic models. It is shown that the density and distribution functions of proposed wrapped Marshall–Olkin circular logistic models are not in closed form. Further, the ideas of Marshall–Olkin stereographic and wrapped circular logistic models are extended to the l-axial case.

On the lines of constructing extemporaneous semicircular / arc models discussed in the previous chapter, here construction of new asymmetric l-axial models using Marshall–Olkin transformation is presented to facilitate data analysis pertaining to asymmetric data, such as face-cleat in a coal seam data (Fisher 1993). This chapter is devoted to constructing stereographic l-axial generalized gamma distribution, Marshall–Olkin stereographic l-axial logistic distribution, Offset l-axial beta distribution, sine skewed l-axial von Mises distribution and wrapped l-axial Marshall–Olkin logistic distribution along with graphs of pdf, cdf, and characteristic functions of the above distributions are shown.

7.2 Stereographic l-Axial Distributions

The stereographic semicircular generalized gamma (SSCGG) distribution and stereographic semicircular versions of Weibull and exponential distributions (Sastry 2016), which are special cases of stereographic semicircular generalized gamma distribution, are extended to the corresponding l-axial distributions, which is applicable to any arc of arbitrary length, say, $^{2\pi}/_{l}$ for $l = 1, 2, \ldots$.

Definition 7.2.1

In the continuous case $g : [0, \frac{2\pi}{l}) \to \mathbb{R}$, $l \in \mathbb{N}$ is the probability density function of a l-axial distribution if and only if g has the following basic properties:

$$g(\theta) \geq 0, \quad \forall \theta$$

$$\int_0^{\frac{2\pi}{l}} g(\theta)\, d\theta = 1$$

$$g(\theta) = g\left(\theta + \frac{2k\pi}{l}\right), \ k \text{ is an integer and } l \in \mathbb{N}$$

7.2.1 Stereographic l-Axial Generalized Gamma Model

A random variable Φ_{sc} on a unit semicircle is said to have stereographic semicircular generalized gamma distribution with shape parameter $c > 0$, index parameter $\alpha > 0$ and scale parameter $\sigma > 0$ denoted by **SSCGG** (c, α, σ), if the probability density for $0 < \theta < \pi, \sigma = \frac{\lambda}{v}, c, \alpha > 0$ is given by

$$g(\theta; c, \alpha, \sigma) = \frac{c}{2\sigma^{\alpha c}\Gamma(\alpha)} \sec^2\left(\frac{\theta}{2}\right)\left(\tan\left(\frac{\theta}{2}\right)\right)^{\alpha c - 1} \exp\left(-\left(\frac{\tan\left(\frac{\theta}{2}\right)}{\sigma}\right)^{c}\right) \quad (7.2.1)$$

By applying the transformation $\theta^* = {}^{2\theta}/_{l}$, $l = 1, 2, \ldots$, on the SSCGG distribution (Figure 7.1), the probability density function of the **stereographic l-axial generalized gamma** (SLAGG) model is given by

$$g(\theta^*) = \frac{cl}{4\sigma^{\alpha c}\Gamma(\alpha)} \sec^2\left(\frac{l\theta^*}{4}\right)\left(\tan\left(\frac{l\theta^*}{4}\right)\right)^{\alpha c - 1} \exp\left(-\left(\frac{\tan\left(\frac{l\theta^*}{4}\right)}{\sigma}\right)^{c}\right) \quad (7.2.2)$$

where $0 < \theta^* < \dfrac{2\pi}{l}$ and $\theta^* = \dfrac{2\theta}{l}$ for $l \in \mathbb{N}$.

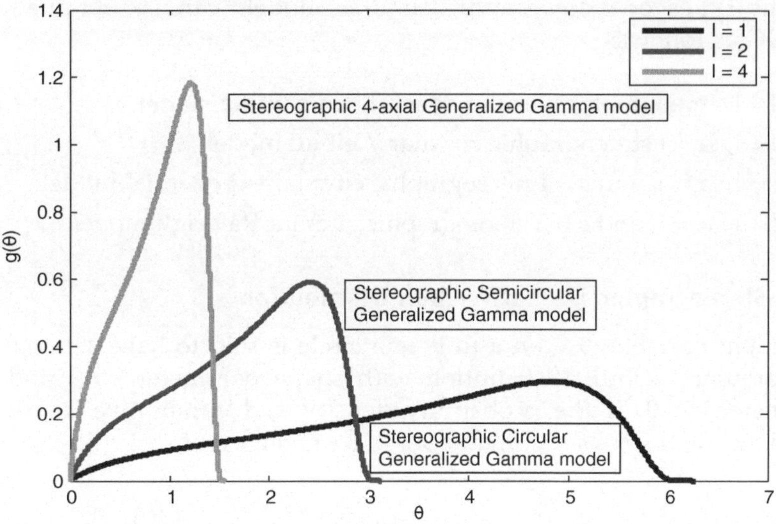

FIGURE 7.1
The graphs of the pdf of the stereographic l-axial generalized gamma model is plotted for $\sigma = 0.5$ and $\alpha = 2.5$.

For $l = 1$ the circular model whose pdf is

$$g(\theta^*) = \frac{c}{4\sigma^{\alpha c}\Gamma(\alpha)}\sec^2\left(\frac{\theta^*}{4}\right)\left(\tan\left(\frac{\theta^*}{4}\right)\right)^{\alpha c-1}\exp\left(-\left(\frac{\tan\left(\frac{\theta^*}{4}\right)}{\sigma}\right)^c\right)$$
$$= \frac{c}{4\sigma^{\alpha c}\Gamma(\alpha)}\sec^2\left(\frac{\theta}{2}\right)\left(\tan\left(\frac{\theta}{2}\right)\right)^{\alpha c-1}\exp\left(-\left(\frac{\tan\left(\frac{\theta}{2}\right)}{\sigma}\right)^c\right) \quad (7.2.3)$$

where $0 < \theta^* < 2\pi \Rightarrow 0 < \theta < \pi$.

For $l = 4$ the 4-axial model

$$g(\theta^*) = \frac{c}{\sigma^{\alpha c}\Gamma(\alpha)}\sec^2(\theta^*)\left(\tan(\theta^*)\right)^{\alpha c-1}\exp\left(-\left(\frac{\tan\theta^*}{\sigma}\right)^c\right) \quad (7.2.4)$$

where $0 < \theta^* < \frac{\pi}{2} \Rightarrow 0 < \theta < \pi$.

Various types of stereographic circular models can be obtained from **SLAGG** as follows:

1. $l=1$ stereographic circular generalized gamma model
2. $l=1,\ \alpha=1$ stereographic circular Weibull model
3. $l=1,\ \alpha=1$, and $c=1$ stereographic circular exponential model
4. $l=1,\ \alpha=1$, and $c=2$ stereographic circular Rayleigh model

7.2.2 Stereographic *l*-Axial Weibull Distribution

A random variable Φ_{SC} on a unit semicircle is said to have stereographic semicircular Weibull distribution with shape parameter $c>0$ and scale parameter $\sigma>0$, if the probability density and cumulative distribution functions for $0<\theta<\pi, c>0$, and $\sigma>0$ are given by

$$g(\theta)=\frac{c}{2\sigma^{c}}\sec^{2}\left(\frac{\theta}{2}\right)\left(\tan\left(\frac{\theta}{2}\right)\right)^{c-1}\exp\left(-\left(\frac{\tan\left(\frac{\theta}{2}\right)}{\sigma}\right)^{c}\right) \tag{7.2.5}$$

$$G(\theta)=1-\exp\left(-\left(\frac{\tan\left(\frac{\theta}{2}\right)}{\sigma}\right)^{c}\right) \tag{7.2.6}$$

The stereographic semicircular Weibull model is extended to the *l*-axial distribution, which is applicable to any arc of arbitrary length, say, $2\pi/l$ for $l=1,2,\ldots$, By applying the transformation $\theta^{*}=2\theta/l, l=1,2,\ldots$, on the stereographic semicircular Weibull distribution, the probability density function of the **stereographic *l*-axial Weibull (SLAW) distribution** (c,σ) is given by

$$g(\theta^{*})=\frac{cl}{4\sigma^{c}}\sec^{2}\left(\frac{l\theta^{*}}{4}\right)\left(\tan\left(\frac{l\theta^{*}}{4}\right)\right)^{c-1}\exp\left(-\left(\frac{\tan\left(\frac{l\theta^{*}}{4}\right)}{\sigma}\right)^{c}\right), \tag{7.2.7}$$

$$0<\theta^{*}<\frac{2\pi}{l},\ \sigma>0,\ c>0, \text{ and } l=1,2,\ldots$$

Case 1: Taking $l=1$ in (7.2.7), the probability density function of **stereographic circular Weibull distribution** and is given by

$$g(\theta^{*})=\frac{c}{4\sigma^{c}}\sec^{2}\left(\frac{\theta^{*}}{4}\right)\left(\tan\left(\frac{\theta^{*}}{4}\right)\right)^{c-1}\exp\left(-\left(\frac{\tan\left(\frac{\theta^{*}}{4}\right)}{\sigma}\right)^{c}\right), \tag{7.2.8}$$

$$0<\theta^{*}<2\pi,\ \sigma>0,\text{ and } c>0$$

Case 2: Taking $l=2$ in (7.2.7) the **stereographic semicircular Weibull distribution** (Phani et al. 2013) is

A random variable Φ_C on a unit circle is said to have stereographic circular Weibull distribution with scale parameters $\sigma>0$, shape parameter $c>0$, and location parameter μ, if its probability density and cumulative distribution functions for $0<\theta,\mu<2\pi$, $c>0$ and $\sigma>0,\theta>\mu$ are given

$$g(\theta)=\frac{c}{4\sigma^{c}}\sec^{2}\left(\frac{\theta-\mu}{4}\right)\left(\tan\left(\frac{\theta-\mu}{4}\right)\right)^{c-1}\exp\left(-\left(\frac{\tan\left(\frac{\theta-\mu}{4}\right)}{\sigma}\right)^{c}\right) \tag{7.2.9}$$

$$G(\theta)=1-\exp\left(-\left(\frac{\tan\left(\frac{\theta-\mu}{4}\right)}{\sigma}\right)^{c}\right) \tag{7.2.10}$$

7.2.3 Stereographic *l*-Axial Exponential Distribution

A random variable Φ_{SC} on a unit semicircle is said to have stereographic semicircular exponential distribution with scale parameter $\sigma>0$, if the probability density and cumulative distribution functions are given by

$$g(\theta)=\frac{1}{2\sigma}\sec^{2}\left(\frac{\theta}{2}\right)\exp\left(-\frac{1}{\sigma}\tan\left(\frac{\theta}{2}\right)\right),\ \text{for } 0<\theta<\pi \ \text{ and } \sigma>0$$

$$G(\theta)=1-\exp\left(-\frac{1}{\sigma}\tan\left(\frac{\theta}{2}\right)\right)$$

The above stereographic semicircular model is extended to the l-axial distribution, which is applicable to any arc of arbitrary length, say, $2\pi/l$ for $l=1,2,....,$ By applying the transformation $\theta^{*}=2\theta/l$, $l=1,2,...,$ on the stereographic semicircular exponential distribution, the probability density function of the **stereographic *l*-axial exponential distribution** is given by

$$g\left(\theta^{*}\right)=\frac{l}{4\sigma}\sec^{2}\left(\frac{l\theta^{*}}{4}\right)\exp\left(-\frac{1}{\sigma}\left(\tan\left(\frac{l\theta^{*}}{4}\right)\right)\right),\ 0<\theta^{*}<\frac{2\pi}{l},\ \sigma>0 \text{ and } l=1,2,.. \tag{7.2.11}$$

Case 1: For $l = 1$ in (7.2.11), the density function of **stereographic circular exponential distribution** is

$$g\left(\theta^*\right) = \frac{1}{4\sigma}\sec^2\left(\frac{\theta^*}{4}\right)\exp\left(-\frac{1}{\sigma}\left(\tan\left(\frac{\theta^*}{4}\right)\right)\right),\ 0 < \theta^* < 2\pi,\ \sigma > 0 \tag{7.2.12}$$

Case 2: For $l = 2$ in (7.2.11), the density function of **stereographic semicircular exponential distribution** (Phani 2013) is

$$g\left(\theta^*\right) = \frac{1}{2\sigma}\sec^2\left(\frac{\theta^*}{2}\right)\exp\left(-\frac{1}{\sigma}\left(\tan\left(\frac{\theta^*}{2}\right)\right)\right),\ 0 < \theta^* < \pi,\ \sigma > 0 \tag{7.2.13}$$

Definition

A random variable Φ_C on a unit circle is said to have stereographic circular exponential distribution with scale parameters $\sigma > 0$ and location parameter μ. If its probability density and cumulative distribution functions are given for $0 < \theta, \mu < 2\pi, \sigma > 0$

$$g(\theta) = \frac{1}{4\sigma}\sec^2\left(\frac{\theta-\mu}{4}\right)\exp\left(-\frac{1}{\sigma}\left(\tan\left(\frac{\theta-\mu}{4}\right)\right)\right) \tag{7.2.14}$$

$$G(\theta) = 1 - \exp\left(-\frac{1}{\sigma}\tan\left(\frac{\theta-\mu}{4}\right)\right) \tag{7.2.15}$$

7.3 Marshall–Olkin Circular Models

Here a new class of asymmetric Marshall–Olkin stereographic and wrapped Marshall–Olkin circular models is introduced (Sastry 2016). The construction of these types of models is demonstrated on a logistic model. It is shown that the density and distribution functions of a proposed wrapped Marshall–Olkin circular logistic model are not in closed form. Further, the ideas of the Marshall–Olkin stereographic and wrapped circular logistic models are extended to the l-axial case, and the above observations are found to be true even in the l-axial case.

7.3.1 Marshall–Olkin Transformation for Circular Data

The idea of Marshall and Olkin (1997) is suitably adopted for circular data and the distribution function of the Marshall–Olkin circular model (denoted as $G_{MO}(\theta)$) in terms of that of the circular model in usual notations by adding the additional tilt parameter $\alpha > 0, -\pi < \theta < \pi$ is defined as

$$G_{MO}(\theta) = \frac{G(\theta)}{\alpha + (1-\alpha)G(\theta)} \tag{7.3.1}$$

The probability density function corresponding to (7.3.1) is

$$g_{MO}(\theta) = \frac{\alpha g(\theta)}{\left(1-(1-\alpha)(1-G(\theta))\right)^2} \tag{7.3.2}$$

NOTE: If $\alpha = 1$, then it can be seen that $G_{MO}(\theta) = G(\theta)$.

7.3.2 Marshall–Olkin Stereographic Circular Logistic Distribution

The probability density and cumulative distribution functions of the stereographic circular logistic distribution (Phani 2013) for $\sigma > 0, -\pi < \theta, \mu < \pi$ are respectively given by

$$g(\theta) = \frac{1}{2\sigma}\sec^2\left(\frac{\theta}{2}\right)\left[1+\exp\left(\frac{-\left(\tan\left(\frac{\theta-\mu}{2}\right)\right)}{\sigma}\right)\right]^{-2}\exp\left(-\frac{\left(\tan\left(\frac{\theta-\mu}{2}\right)\right)}{\sigma}\right) \tag{7.3.3}$$

$$G(\theta) = \left[1+\exp\left(\frac{-\left(\tan\left(\frac{\theta-\mu}{2}\right)\right)}{\sigma}\right)\right]^{-1} \tag{7.3.4}$$

NOTE: Properties of the stereographic circular logistic model are

1. The stereographic logistic model is symmetric about $\mu = 0$.
2. The stereographic logistic distribution is **unimodal** if $\sigma < 0.5$ and **bimodal** if $\sigma > 0.5$.
3. $\frac{1}{\sigma}$ is a parameter that measures the concentration toward the mean direction.

By applying the Marshall–Olkin transformation for circular data on the stereographic circular logistic distribution, a new family of stereographic circular logistic distribution is obtained and is called the Marshall–Olkin stereographic circular logistic distribution.

Definition 7.3.1

A random variable Φ_C on a unit circle is said to have the Marshall–Olkin stereographic circular logistic (MOSCLG) distribution with location parameter μ, scale parameter $\sigma > 0$ and tilt parameter $\alpha > 0$ denoted by **MOSCLG** (μ, σ, α), if the probability density and cumulative distribution functions for $\sigma, \alpha > 0$ and $-\pi < \theta, \mu < \pi$ are respectively given by

$$g_{MO}(\theta) = \frac{\alpha}{2\sigma}\sec^2\left(\frac{\theta}{2}\right)\left[1+\alpha\exp\left(\frac{-\left(\tan\left(\frac{\theta-\mu}{2}\right)\right)}{\sigma}\right)\right]^{-2}\exp\left(-\left(\frac{\tan\left(\frac{\theta-\mu}{2}\right)}{\sigma}\right)\right) \tag{7.3.5}$$

$$G_{MO}(\theta) = \left[1+\alpha\exp\left(\frac{-\left(\tan\left(\frac{\theta-\mu}{2}\right)\right)}{\sigma}\right)\right]^{-1} \tag{7.3.6}$$

Graphs of probability density function and cdf of the MOSCLG distribution for various values of $\sigma = 1.5$, α, and $\mu = 0$ are presented (Figures 7.2 through 7.4).

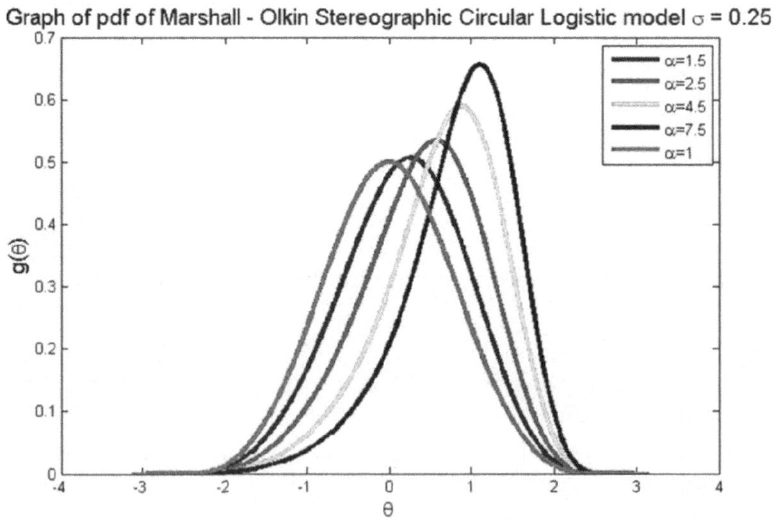

FIGURE 7.2
The graphs of the pdf of Marshall–Olkin stereographic circular logistic distribution $\sigma = 0.25$.

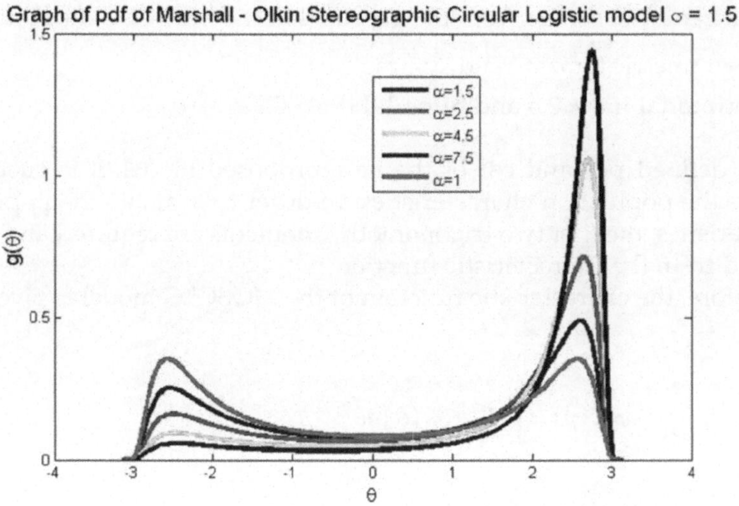

FIGURE 7.3
The graphs of the pdf of Marshall–Olkin stereographic circular logistic distribution $s = 1.5$.

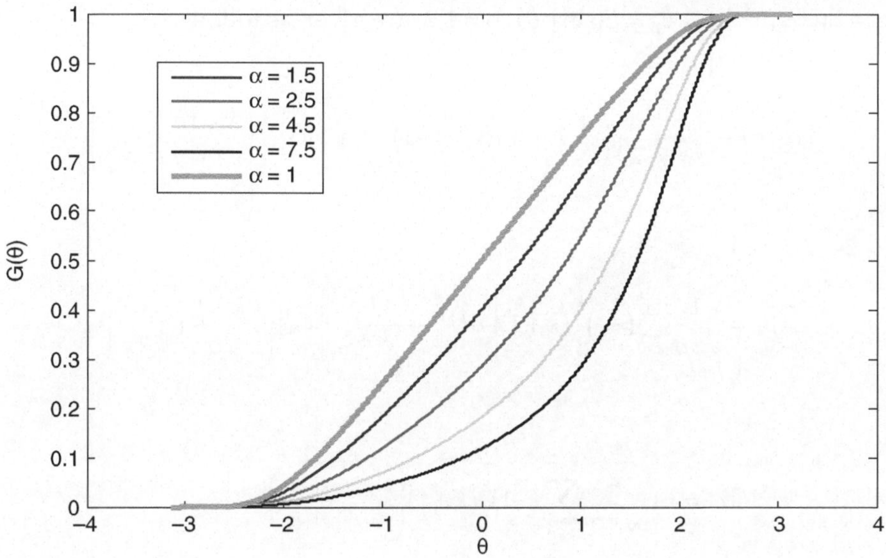

FIGURE 7.4
The graphs of the cdf of the Marshall–Olkin stereographic circular logistic distribution.

NOTE: It can be observed from the graphs that the MOSCLG distribution is

1. Asymmetric for $\alpha \neq 1$ and symmetric for $\alpha = 1$
2. **Unimodal** if $\sigma < 0.5$ and **bimodal** if $\sigma > 0.5$

Having defined pdf and cdf of the new proposed model, it is customary to study the population characteristics. In order to evaluate the population characteristics, the first two trigonometric moments are required and can be obtained from the characteristic function.

Therefore, the characteristic function of the MOSCLG model is given by

$$\varphi_{X_S}(p) = \int_{-\pi}^{\pi} e^{ip\theta} g_{MO}(\theta) d\theta$$
$$= \int_{-\infty}^{\infty} e^{ip\left(2\tan^{-1}(x)\right)} \frac{\alpha}{\sigma}\left[1+\alpha e^{-\left(\frac{x}{\sigma}\right)}\right]^{-2} e^{-\left(\frac{x}{\sigma}\right)} dx$$

Theorem 7.3.1

Under the pdf of the MOSCLG distribution with $\mu = 0$, the first two $\alpha_p = E(\cos p\theta)$ and $\beta_p = E(\sin p\theta)$, $p = 1,2$, are given as follows:

$$\alpha_1 = 1 - \frac{1}{\sqrt{\pi}\sigma} \sum_{k=0}^{\infty} \left[\left(\frac{1}{\alpha}\right)^{k+1} + \alpha^{k+1}\right] (-1)^k (k+1) G_{13}^{31}\left(\left(\frac{k+1}{2\sigma}\right)^2 \middle| \begin{matrix} \frac{-1}{2} \\ \frac{-1}{2}, 0, \frac{1}{2} \end{matrix}\right)$$

$$\beta_1 = \frac{1}{\sqrt{\pi}\sigma} \sum_{k=0}^{\infty} (-1)^k (k+1) \left(\alpha^{k+1} - \frac{1}{\alpha^{k+1}}\right) G_{13}^{31}\left(\left(\frac{k+1}{2\sigma}\right)^2 \middle| \begin{matrix} 0 \\ 0, 0, \frac{1}{2} \end{matrix}\right)$$

$$\alpha_2 = 1 + \frac{4}{\sqrt{\pi}\sigma} \sum_{k=0}^{\infty} (-1)^k (k+1) \left(\alpha^{k+1} + \left(\frac{1}{\alpha}\right)^{k+1}\right)$$
$$\left(G_{13}^{31}\left(\left(\frac{k+1}{2\sigma}\right)^2 \middle| \begin{matrix} -\frac{3}{2} \\ -\frac{1}{2}, 0, \frac{1}{2} \end{matrix}\right) + G_{13}^{31}\left(\left(\frac{k+1}{2\sigma}\right)^2 \middle| \begin{matrix} -\frac{1}{2} \\ -\frac{1}{2}, 0, \frac{1}{2} \end{matrix}\right)\right)$$

$$\beta_2 = \frac{2}{\sqrt{\pi}\sigma}\sum_{k=0}^{\infty}(-1)^k(k+1)\left(\alpha^{k+1}-\left(\frac{1}{\alpha}\right)^{k+1}\right)$$

$$\left[G_{13}^{31}\left(\left(\frac{k+1}{2\sigma}\right)^2\middle|\begin{matrix}0\\0,0,\frac{1}{2}\end{matrix}\right)-2G_{13}^{31}\left(\left(\frac{k+1}{2\sigma}\right)^2\middle|\begin{matrix}-1\\0,0,\frac{1}{2}\end{matrix}\right)\right]$$

where

$$\int_0^{\infty} x^{2\nu-1}\left(u^2+x^2\right)^{Q-1}e^{-\mu x}dx = \frac{u^{2\nu+2Q-2}}{2\sqrt{\pi}\Gamma(1-Q)}G_{13}^{31}\left(\frac{\mu^2u^2}{4}\middle|\begin{matrix}1-\nu\\1-Q-\nu,0,\frac{1}{2}\end{matrix}\right) \tag{7.3.7}$$

for $|\arg u\pi| < \frac{\pi}{2}$, $\mathrm{Re}\,\mu > 0$ and $\mathrm{Re}\,\nu > 0$ and $G_{13}^{31}\left(\frac{\mu^2u^2}{4}\middle|\begin{matrix}1-\nu\\1-Q-\nu,0,\frac{1}{2}\end{matrix}\right)$ is called as Meijer's G-function (Gradshteyn and Ryzhik 2007, formula no. 3.389.2) (Figure 7.5).

From the population characteristics, it can be observed that with increasing value of parameter α and fixed value of σ, the circular variance gradually decreases, the distribution is negatively skewed and shifted from leptokurtic to platykurtic.

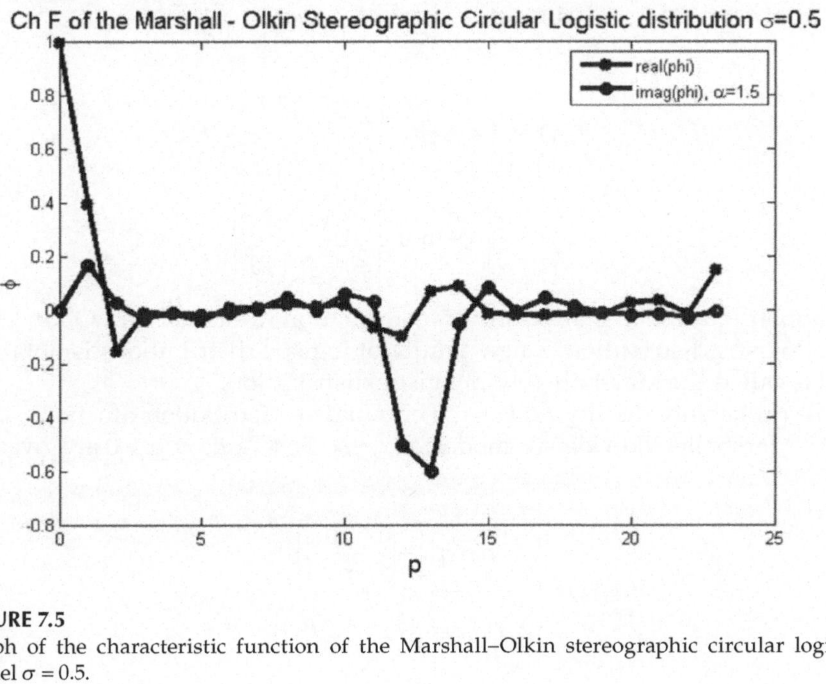

FIGURE 7.5
Graph of the characteristic function of the Marshall–Olkin stereographic circular logistic model $\sigma = 0.5$.

7.3.3 Wrapped Marshall–Olkin Logistic Distribution

In this section, the wrapped Marshall–Olkin logistic distribution is derived as follows:

- The transformation of Marshall and Olkin (1997) is applied on a logistic distribution and the resultant family of asymmetric distributions thus obtained is named by us as Marshall–Olkin logistic distributions.
- Wrapping is applied on the Marshall–Olkin logistic distribution to derive a new family of asymmetric circular models called wrapped Marshall–Olkin logistic distributions.

In order to derive trigonometric moments of a wrapped Marshall–Olkin logistic distribution, the characteristic function of the Marshall–Olkin logistic distribution is derived. Using Proposition 2.1 (Jammalamadaka and Sen Gupta 2001), the characteristic function of the proposed circular model is obtained. The technical details are presented as follows:

The pdf $f(x)$ and cdf $F(x)$ of the logistic distribution for $x, \mu \in, x \geq \mu$ and $\sigma > 0$ are given by

$$\begin{aligned} f(x) &= \frac{1}{4\sigma}\sec h^2\left(\frac{x-\mu}{2\sigma}\right), \\ &= \frac{1}{\sigma}\left[1+\exp\left\{\frac{-(x-\mu)}{\sigma}\right\}\right]^{-2}\exp\left\{\frac{-(x-\mu)}{\sigma}\right\} \end{aligned} \tag{7.3.8}$$

$$\begin{aligned} F(x) &= \left[1+\exp\left\{\frac{-(x-\mu)}{\sigma}\right\}\right]^{-1} \\ &= \frac{1}{2}\left[1+\tanh\left\{\frac{1}{2}\left(\frac{x-\mu}{\sigma}\right)\right\}\right] \end{aligned} \tag{7.3.9}$$

By applying the Marshall–Olkin transformation (Marshall and Olkin 1997) on a logistic distribution, a new family of logistic distributions is obtained and is called the Marshall–Olkin logistic distribution.

The probability density $f_{MO}(x)$ and cumulative distribution function $F_{MO}(x)$ of the Marshall–Olkin logistic model for $x,\ \mu \in \mathbb{R},\ x\ \geq \mu,\ \sigma, \alpha > 0$ are given by

$$f_{MO}(x) = \frac{\alpha \exp\left\{-\left(\frac{x-\mu}{\sigma}\right)\right\}}{\sigma\left(1+\alpha\exp\left\{-\left(\frac{x-\mu}{\sigma}\right)\right\}\right)^2}$$

$$F_{MO}(x)=\left[1+\alpha\exp\left\{\frac{-(x-\mu)}{\sigma}\right\}\right]^{-1}$$

The characteristic function of the Marshal–Olkin logistic distribution for $x \in (-\infty,\infty), \alpha,\sigma > 0$ and $t \in \mathbb{R}$ is

$$\phi_X(t)=\int_{-\infty}^{\infty} e^{itx} f(x)dx$$

$$=\frac{\alpha}{\sigma}\int_{-\infty}^{\infty} e^{itx}\exp\left\{-\left(\frac{x-\mu}{\sigma}\right)\right\}\left(1+\alpha\exp\left\{-\left(\frac{x-\mu}{\sigma}\right)\right\}\right)^{-2} dx,$$

$$=e^{it\mu}\alpha^{it\sigma}\frac{\pi\sigma t}{\sinh\pi\sigma t}, \text{ for } t\in R$$

$$\phi_X(t)=\alpha^{it\sigma}\left(\frac{e^{it\mu}\pi\sigma t}{\sinh\pi\sigma t}\right), \text{ for } t\in R$$

The characteristic function of the wrapped Marshall–Olkin logistic model for $p \in Z$ is given by

$$\varphi_{X_W}(p)=\alpha^{ip\sigma}\left[\cos p\mu\frac{\pi\sigma p}{\sinh\pi\sigma p}+i\sin p\mu\frac{\pi\sigma p}{\sinh\pi\sigma p}\right]$$

$$=\frac{\pi\sigma p}{\sinh\pi\sigma p}\cos p(\mu+\sigma\log\alpha)+i\frac{\pi\sigma p}{\sinh\pi\sigma p}\sin p(\mu+\sigma\log\alpha)$$

$$\varphi_{X_W}(p)=\alpha_p+i\beta_p,$$

where $\alpha_p=\frac{\pi\sigma p}{\sinh\pi\sigma p}\cos p(\mu+\sigma\log\alpha)$ and $\beta_p=\frac{\pi\sigma p}{\sinh\pi\sigma p}\sin p(\mu+\sigma\log\alpha)$ are the trigonometric moments.

Using these trigonometric moments, Fourier representation for the density function of the wrapped circular random variable X_W is given by

$$g(\theta;\mu,\sigma,\alpha)=\frac{1}{2\pi}\left[1+2\sum_{p=1}^{\infty}\{\alpha_p\cos p(\theta-\mu)+\beta_p\sin p(\theta-\mu)\}\right]$$

$$=\frac{1}{2\pi}\left[1+2\pi\sigma\sum_{p=1}^{\infty}\left\{\frac{p}{\sinh\pi\sigma p}\{\cos p(\theta-(2\mu+\sigma\log\alpha))\}\right\}\right]$$

By applying wrapping on the Marshall–Olkin logistic distribution, the probability density and cumulative distribution functions of the wrapped Marshall–Olkin logistic distribution for $\theta > \mu$ and $\theta, \mu \in [0,2\pi), \sigma > 0, \alpha > 0$ are respectively given by (Figures 7.6 and 7.7)

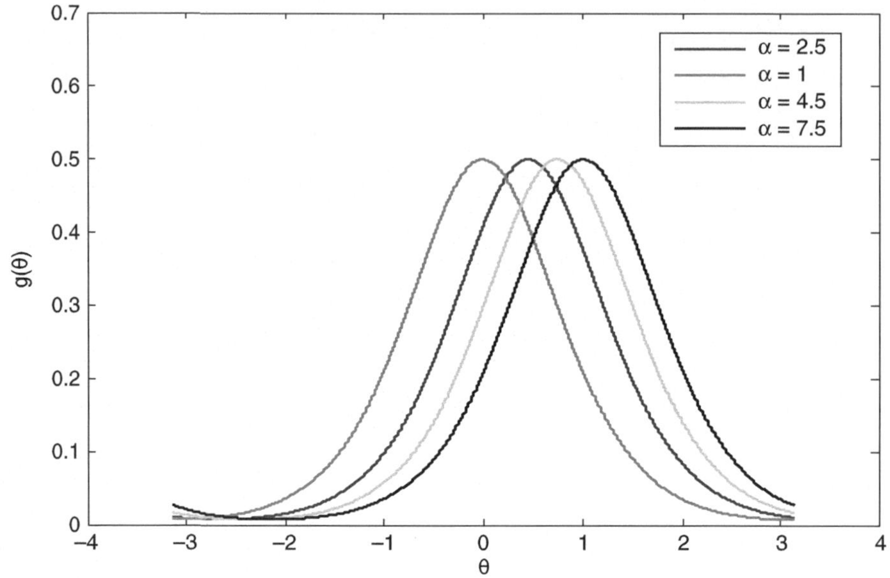

FIGURE 7.6
The graphs of the pdf of wrapped Marshall–Olkin logistic distribution for $\sigma = 0.5$.

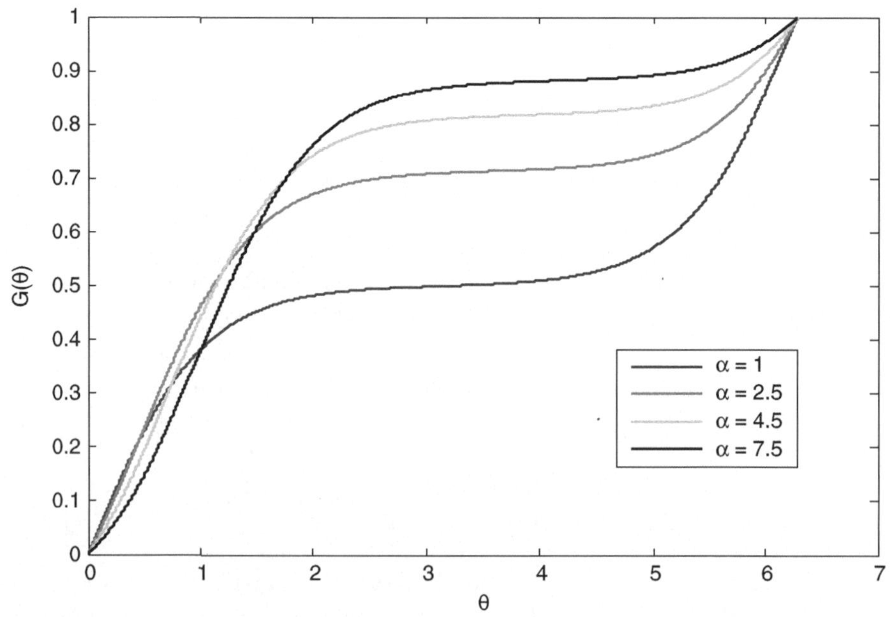

FIGURE 7.7
The graphs of the cdf of wrapped Marshall–Olkin logistic distribution for $\sigma = 0.5$.

$$g_{MO}(\theta) = \sum_{k=-\infty}^{\infty} \frac{\alpha}{\sigma}\left[1+\alpha \exp\left\{\frac{-(\theta+2k\pi-\mu)}{\sigma}\right\}\right]^{-2} \exp\left\{\frac{-(\theta+2k\pi-\mu)}{\sigma}\right\}$$
$$= \sum_{k=-\infty}^{\infty} \frac{1}{4\sigma} \operatorname{sec}h^2\left(\frac{(\theta+2k\pi-\mu)}{2\sigma}\right) \tag{7.3.10}$$

$$G_{MO}(\theta) = \sum_{k=-\infty}^{\infty}\left[1+\alpha \exp\left\{\frac{-(\theta+2k\pi-\mu)}{\sigma}\right\}\right]^{-1} - \left[1+\alpha \exp\left\{\frac{-(2k\pi-\mu)}{\sigma}\right\}\right]^{-1} \tag{7.3.11}$$

Definition 7.3.2

A random variable Φ_C on a unit circle is said to have the wrapped circular Marshall–Olkin logistic distribution with location parameter μ, scale parameter $\sigma > 0$ and tilt parameter $\alpha > 0$ and is denoted by WCMOLG (μ,σ,α), if the probability density and cumulative distribution functions are respectively given in (7.3.10) and (7.3.11).

The population characteristics of the WCMOLG distribution are evaluated. From the population characteristics, it can be observed that with an increasing value of parameter α and fixed value of σ, the circular variance gradually decreases and the distribution is positively skewed and shifted from leptokurtic to platykurtic.

7.4 Marshall–Olkin Stereographic *l*-Axial Logistic Distribution

The Marshall–Olkin stereographic circular logistic model is generalized to the *l*-axial distribution, which is applicable to any arc of arbitrary length, say, $\frac{2\pi}{l}$ for $l=1,2,....$ By applying the transformation $\theta^* = \theta/l$, $l=1,2,...,$ on the Marshall–Olkin stereographic circular logistic distribution, the probability density function of **Marshall–Olkin stereographic *l*-axial logistic (MOSLL) distribution** (α,σ) **for** $\sigma,\alpha > 0, -\frac{\pi}{l} < \theta^* < \frac{\pi}{l}$ is given by

$$g_{MO}(\theta^*) = \frac{\alpha l}{2\sigma} \sec^2\left(\frac{l\theta^*}{2}\right)\left[1+\alpha \exp\left(-\frac{\tan\left(\frac{l\theta^*}{2}\right)}{\sigma}\right)\right]^{-2} \exp\left(-\frac{\tan\left(\frac{l\theta^*}{2}\right)}{\sigma}\right) \tag{7.4.1}$$

$$G_{MO}(\theta^*) = \left[1+\alpha \exp\left(\frac{-\left(\tan\left(\frac{l\theta^*}{2}\right)\right)}{\sigma}\right)\right]^{-1} \tag{7.4.2}$$

Case 1: When $l=1$, the probability density and cumulative distribution functions are the same as that of Marshall–Olkin stereographic circular logistic distribution for $\sigma,\alpha>0,\ -\pi<\theta^*<\pi$ given by

$$g_{MO}(\theta^*)=\frac{\alpha}{2\sigma}\sec^2\left(\frac{\theta^*}{2}\right)\left[1+\alpha\exp\left(-\frac{\tan\left(\frac{\theta^*}{2}\right)}{\sigma}\right)\right]^{-2}\exp\left(-\frac{\tan\left(\frac{\theta^*}{2}\right)}{\sigma}\right) \tag{7.4.3}$$

$$G_{MO}(\theta^*)=\left[1+\alpha\exp\left(\frac{-\left(\tan\left(\frac{\theta^*}{2}\right)\right)}{\sigma}\right)\right]^{-1} \tag{7.4.4}$$

Case 2: When $l=2$, the probability density and cumulative distribution functions of the Marshall–Olkin stereographic semicircular logistic distribution for $\sigma,\alpha>0,\ -\frac{\pi}{2}<\theta^*<\frac{\pi}{2}$ are

$$g_{MO}\left(\theta^*\right)=\frac{\alpha}{\sigma}\sec^2\left(\theta^*\right)\left[1+\alpha\exp\left(-\frac{\tan\left(\theta^*\right)}{\sigma}\right)\right]^{-2}\exp\left(-\frac{\tan\left(\theta^*\right)}{\sigma}\right) \tag{7.4.5}$$

$$G_{MO}(\theta^*)=\left[1+\alpha\exp\left(\frac{-\left(\tan\left(\theta^*\right)\right)}{\sigma}\right)\right]^{-1} \tag{7.4.6}$$

7.5 Wrapped l-Axial Marshall–Olkin Logistic Distribution

The pdf and the cdf of the wrapped Marshall–Olkin logistic distribution for $0\le\theta,\mu<2\pi,\alpha>0,\sigma>0$ are

$$g_{MO}(\theta)=\sum_{k=-\infty}^{\infty}\frac{\alpha}{\sigma}\left[1+\alpha\exp\left(\frac{-(\theta+2k\pi-\mu)}{\sigma}\right)\right]^{-2}\exp\left(\frac{-(\theta+2k\pi-\mu)}{\sigma}\right)$$

$$G_{MO}(\theta)=\sum_{k=-\infty}^{\infty}\left(\left[1+\alpha\exp\left(\frac{-(\theta+2k\pi-\mu)}{\sigma}\right)\right]^{-1}-\left[1+\alpha\exp\left(\frac{-(2k\pi-\mu)}{\sigma}\right)\right]^{-1}\right)$$

This is extended to the wrapped l-axial Marshall–Olkin logistic distribution.

The pdf and the cdf of the wrapped l-axial Marshall–Olkin logistic distribution for $\theta^* = \theta/l, 0 \le \theta < 2\pi, 0 \le \theta^*, \mu < \frac{2\pi}{l}, \alpha, \sigma > 0$ for $l = 1,2,\ldots$ are given by

$$g_{MO}(\theta^*) = \sum_{k=-\infty}^{\infty} \frac{l\alpha}{\sigma}\left[1+\alpha\exp\left(\frac{-(l\theta^* + 2k\pi - \mu)}{\sigma}\right)\right]^{-2} \exp\left(\frac{-(l\theta^* + 2k\pi - \mu)}{\sigma}\right) \quad (7.5.1)$$

$$G_{MO}(\theta^*) = \sum_{k=-\infty}^{\infty}\left(\left[1+\alpha\exp\left(\frac{-(l\theta^* + 2k\pi - \mu)}{\sigma}\right)\right]^{-1} - \left[1+\alpha\exp\left(\frac{-(2k\pi - \mu)}{\sigma}\right)\right]^{-1}\right) \quad (7.5.2)$$

Case 1: For $l = 1$ the pdf and the cdf of the wrapped circular Marshall–Olkin logistic distribution for $0 \le \theta^* < 2\pi$ are given by

$$g_{MO}(\theta^*) = \sum_{k=-\infty}^{\infty} \frac{\alpha}{\sigma}\left[1+\alpha\exp\left(\frac{-(\theta^* + 2k\pi - \mu)}{\sigma}\right)\right]^{-2} \exp\left(\frac{-(\theta^* + 2k\pi - \mu)}{\sigma}\right) \quad (7.5.3)$$

$$G_{MO}(\theta^*) = \sum_{k=-\infty}^{\infty}\left(\left[1+\alpha\exp\left(\frac{-(\theta^* + 2k\pi - \mu)}{\sigma}\right)\right]^{-1} - \left[1+\alpha\exp\left(\frac{-(2k\pi - \mu)}{\sigma}\right)\right]^{-1}\right) \quad (7.5.4)$$

Case 2: For $l = 2$ the pdf and the cdf of the wrapped semicircular Marshall–Olkin logistic distribution for $0 \le \theta^* < \pi$ are given by

$$g_{MO}\left(\theta^*\right) = \sum_{k=-\infty}^{\infty} \frac{2\alpha}{\sigma}\left[1+\alpha\exp\left(\frac{-\left(2\theta^* + 2k\pi - \mu\right)}{\sigma}\right)\right]^{-2} \exp\left(\frac{-\left(2\theta^* + 2k\pi - \mu\right)}{\sigma}\right) \quad (7.5.5)$$

$$G_{MO}\left(\theta^*\right) = \sum_{k=-\infty}^{\infty}\left(\left[1+\alpha\exp\left(\frac{-\left(2\theta^* + 2k\pi - \mu\right)}{\sigma}\right)\right]^{-1} - \left[1+\alpha\exp\left(\frac{-(2k\pi - \mu)}{\sigma}\right)\right]^{-1}\right) \quad (7.5.6)$$

7.6 Other Skewed *l*-Axial Models

Girija et al. (2014) obtained the arc offset beta (**AOB**) distribution directly by applying offsetting on the bivariate beta distribution (Balakrishnan and Chin-Diew 2008, p. 374), without imposing any restriction on a circular random variable as a natural phenomenon (Sastry 2016).

7.6.1 Offset *l*-Axial Beta Model

The pdf $g(\theta)$ and cdf $G(\theta)$ of the AOB model with parameters $a, b > 0$ and $0 \le \theta < \frac{\pi}{2}$ are respectively given by

$$g(\theta) = \frac{1}{\beta(a,b)} \frac{\cos^{(a-1)}\theta \, \sin^{(b-1)}\theta}{(\cos\theta + \sin\theta)^{a+b}} \tag{7.6.1}$$

$$G(\theta) = 1 - \frac{\beta_p(a,b)}{\beta(a,b)} \text{ where } p = \frac{1}{1+\tan\theta} \tag{7.6.2}$$

This is extended to offset *l*-axial beta distribution.

The pdf and the cdf of the offset *l*-axial beta distribution for $0 \le \theta^* < \frac{2\pi}{l}$, $a,b > 0$ and $\theta^* = \frac{4\theta}{l}, 0 \le \theta < \frac{\pi}{2}$ are given by

$$g(\theta^*) = \frac{l}{4\beta(a,b)} \frac{\cos^{(a-1)}\left(\frac{l\theta^*}{4}\right)\sin^{(b-1)}\left(\frac{l\theta^*}{4}\right)}{\left(\cos\left(\frac{l\theta^*}{4}\right) + \sin\left(\frac{l\theta^*}{4}\right)\right)^{a+b}} \tag{7.6.3}$$

$$G(\theta^*) = 1 - \frac{\beta_p(a,b)}{\beta(a,b)} \text{ where } p = \frac{1}{1+\tan\left(\frac{l\theta^*}{4}\right)} \tag{7.6.4}$$

The pdf and the cdf of the offset semicircular beta model for $l = 2$ and $0 \le \theta^* < \pi$, $a,b > 0$ are given by

$$g(\theta^*) = \frac{l}{2\beta(a,b)} \frac{\cos^{(a-1)}\left(\frac{\theta^*}{2}\right)\sin^{(b-1)}\left(\frac{\theta^*}{2}\right)}{\left(\cos\left(\frac{\theta^*}{2}\right) + \sin\left(\frac{\theta^*}{2}\right)\right)^{a+b}} \tag{7.6.5}$$

$$G(\theta^*) = 1 - \frac{\beta_p(a,b)}{\beta(a,b)} \quad \text{where } p = \frac{1}{1+\tan\left(\frac{\theta^*}{2}\right)} \tag{7.6.6}$$

7.6.2 Sine Skewed *l*-Axial von Mises Model

Umbach and Rao (2012) obtained an asymmetric circular model called sine skewed von Mises model by skewing the von Mises model and the pdf and the cdf, respectively, for $-\pi \le \theta < \pi, -1 \le \lambda \le 1, \kappa > 0,$ given by them are

$$g(\theta) = \frac{e^{k\cos(\theta-\mu)}}{2\pi I_0(k)}\left(1+\lambda\sin(\theta-\mu)\right) \tag{7.6.7}$$

$$G(\theta) = \frac{\theta I_0(\kappa) + 2\sum_{j=1}^{\infty}\frac{I_j(\kappa)\sin(j\theta)}{j}}{2\pi I_0(\kappa)} + \frac{\lambda}{2\pi\kappa I_0(\kappa)}\left(e^{-\kappa} - e^{\kappa\cos\theta}\right) \tag{7.6.8}$$

This is extended to the sine skewed *l*-axial von Mises distribution.

The pdf and the cdf of the sine skewed *l*-axial von Mises distribution for $-\frac{\pi}{l} \le \theta^* < \frac{\pi}{l}, -1 \le \lambda \le 1, \kappa > 0,$ are given by

$$g\left(\theta^*\right) = \frac{le^{k\cos\left(l\theta^*\right)}}{2\pi I_0(k)}\left(1+\lambda\sin\left(l\theta^*\right)\right) \tag{7.6.9}$$

$$G\left(\theta^*\right) = \frac{l\theta^* I_0(\kappa) + 2\sum_{j=1}^{\infty}\frac{I_j(\kappa)\sin\left(jl\theta^*\right)}{j}}{2\pi I_0(\kappa)} + \frac{\lambda}{2\pi\kappa I_0(\kappa)}\left(e^{-\kappa} - e^{\kappa\cos l\theta^*}\right) \tag{7.6.10}$$

The pdf and the cdf of the sine skewed semicircular von Mises distribution for $l = 2, -\frac{\pi}{2} \le \theta^* < \frac{\pi}{2}, -1 \le \lambda \le 1, \kappa > 0$ are given by

$$g\left(\theta^*\right) = \frac{e^{k\cos\left(2\theta^*-\mu\right)}}{\pi I_0(k)}\left(1+\lambda\sin\left(2\theta^*-\mu\right)\right) \tag{7.6.11}$$

$$G\left(\theta^*\right) = \frac{\theta^* I_0(\kappa) + \sum_{j=1}^{\infty}\frac{I_j(\kappa)\sin\left(j2\theta^*\right)}{j}}{\pi I_0(\kappa)} + \frac{\lambda}{2\pi\kappa I_0(\kappa)}\left(e^{-\kappa} - e^{\kappa\cos 2\theta^*}\right) \tag{7.6.12}$$

References

Balakrishnan, N. and Chin-Diew, L.A.I. 2008. *Continuous Bivariate Distributions*, Springer, New York.

Girija, S.V.S., Radhika, A.J.V. and Dattatreya Rao, A.V. 2014. On Offset l-Arc Models, *Mathematics and Statistics*, 2(3), 127–136.

Gradshteyn, I.S. and Ryzhik, I.M. 2007. *Table of Integrals, Series and Products*, 7th ed., Academic Press, London, UK.

Fisher, N.I. 1993. *Statistical Analysis of Circular Data*, Cambridge University Press, Cambridge, UK.

Jammalamadaka, S.R. and Sen Gupta, A. 2001. *Topics in Circular Statistics*, World Scientific Press, Singapore.

Marshall, A.W. and Olkin, I. 1997. A New Method for Adding a Parameter to a Family of Distributions with Application to the Exponential and Weibull Families, *Biometrika*, 84(3), 641–652.

Phani, Y. 2013. On Stereographic Circular and Semicircular Models, PhD dissertation, Acharya Nagarjuna University, Guntur, India.

Phani, Y., Girija, S.V.S. and Dattatreya Rao, A.V. 2013. On Construction of Stereographic Semicircular Models, *Journal of Applied Probability and Statistics*, 8(1), 75–90.

Sastry, C.H.V. 2016. On *l*-Axial Models. PhD dissertation, Acharya Nagarjuna University, Guntur, India.

Umbach, D. and Rao, J.S. 2012. On Introducing Asymmetry into Circular Distributions, *PJSOR*, 8, 531–535.

8

Choice of Angular Models

8.1 Introduction

After the construction of a new distribution, the next logical step would be to fit these distributions as a model to a live data. This is for finding out how best these new models will fit to a given live data set, and if so, what would be the relative performance with respect to the other good-fit models in the class.

This chapter is devoted to fit wrapped exponentiated inverted Weibull distribution and wrapped new Weibull–Pareto distribution to a live circular data set (Jammalamadaka and Sen Gupta 2001, p. 5) and stereographic semicircular exponentiated inverted Weibull distribution and stereographic semicircular new Weibull–Pareto distribution to a live semicircular data set (Fisher 1993, p. 242).

8.2 Live Data Sets

The details of the two data sets selected are discussed.

8.2.1 Live Data Set 1: Movements of Turtles (Circular)

This circular data set pertains to the movement of turtles and is considered for the two circular models as the data spans in $(0, 2\pi)$. The orientations of 76 turtles after laying eggs are recorded. The data of 76 angular observations in degrees are furnished hereunder (Figure 8.1).

8	9	13	13	14	18	22	27	30	34	38	38	40
44	45	47	48	48	48	48	50	53	56	57	58	58
61	63	64	64	64	65	65	68	70	73	78	78	7
83	83	88	88	88	90	92	92	93	95	96	98	100
103	106	113	118	138	153	153	155	204	215	223	226	237
238	243	244	250	251	257	268	285	319	343	350		

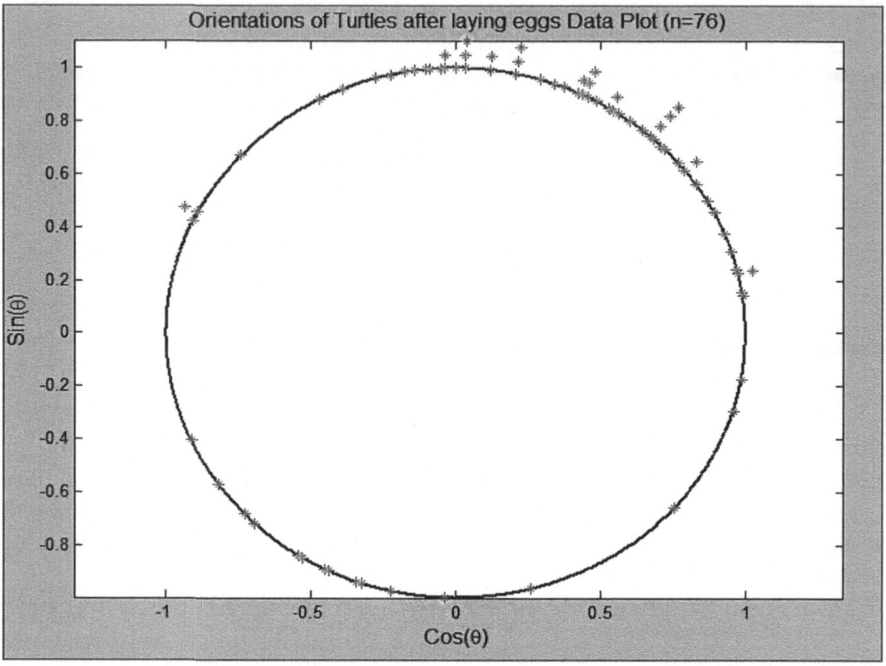

FIGURE 8.1
Data plot for movement of turtle data.

8.2.2 Live Data Set 2: Long-Axis Orientation of Feldspar Laths (Semicircular)

This semicircular data set pertains to the long-axis orientation of feldspar laths and is considered for the two semicircular models as the data ranges in $(0, \pi)$. The data furnished hereunder consists of 60 angular observations in degrees (Figure 8.2).

1	1	2	2	3	8	9	12	16	17
19	23	28	28	34	34	35	36	36	37
41	45	49	50	51	53	58	68	69	70
72	72	76	78	80	85	97	97	99	101
105	121	125	126	133	141	143	149	152	156
160	163	167	168	170	171	172	174	175	176

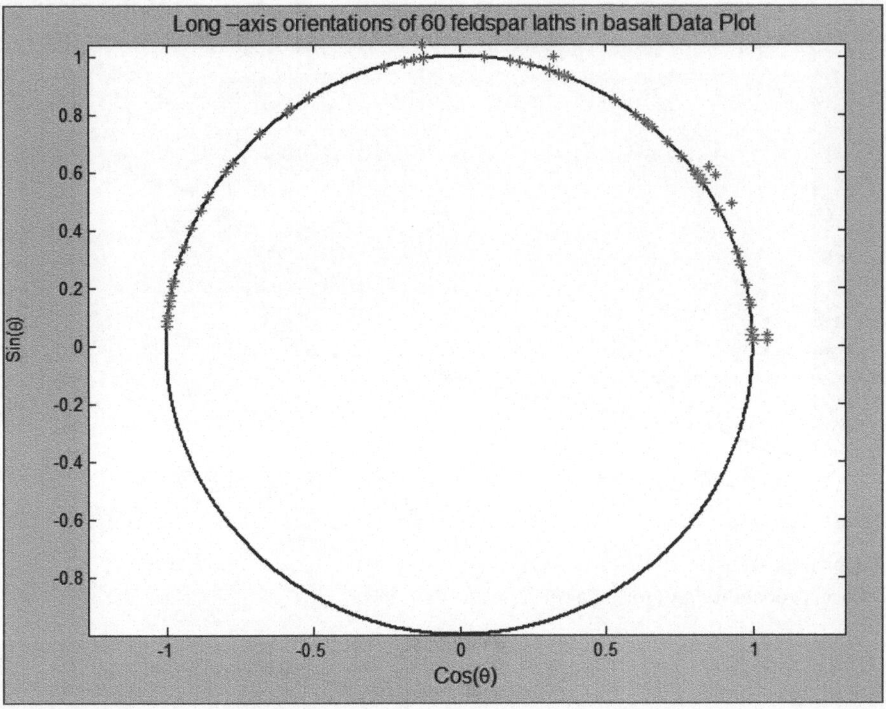

FIGURE 8.2
Data plot for long-axis orientation of feldspar laths.

8.2.3 Uniform Probability Plot

Uniform probability plots are used to know whether data sets are drawn from the uniform distribution. This uniform probability plot can be constructed by adding 0 and 2π to the circular data set considered, arranging the observations in increasing order as $0 \le \theta_{(1)} \le \theta_{(2)} \le ... \le \theta_{(n)} \le 2\pi$ with respect to some initial direction, and then $\theta_i / 2\pi$ is plotted against $i/(n+1)$ for $i = 1,2,...,n+1$. In case of a semicircular data set, π in place 2π is added to the data set, and θ_i / π is taken for plotting against $i/(n+1)$. If the random sample $\theta_1, \theta_2, ..., \theta_n$ is uniformly distributed, then the points should lie near a straight line of slope 45° passing through the origin. The probability plots for data sets 1 and 2 are furnished in Figures 8.3 and 8.4.

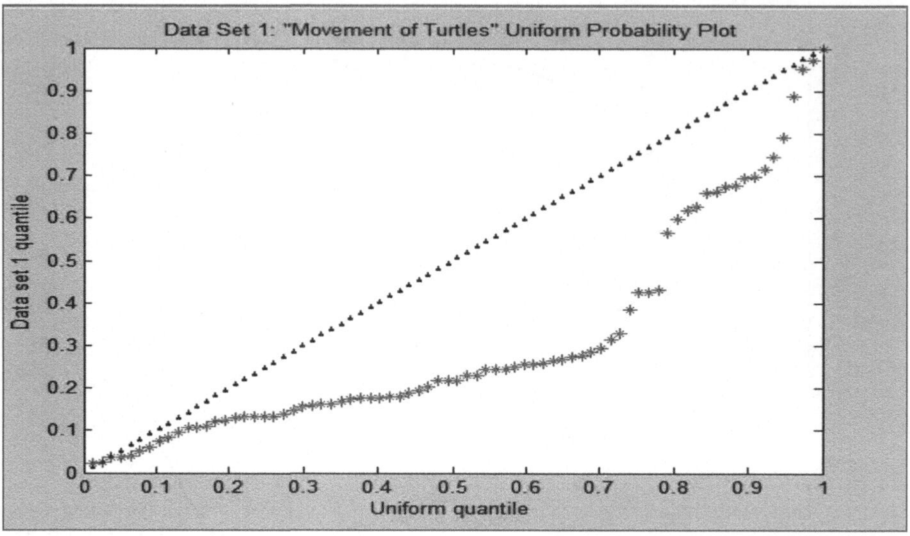

FIGURE 8.3
Uniform probability plot for data set 1.

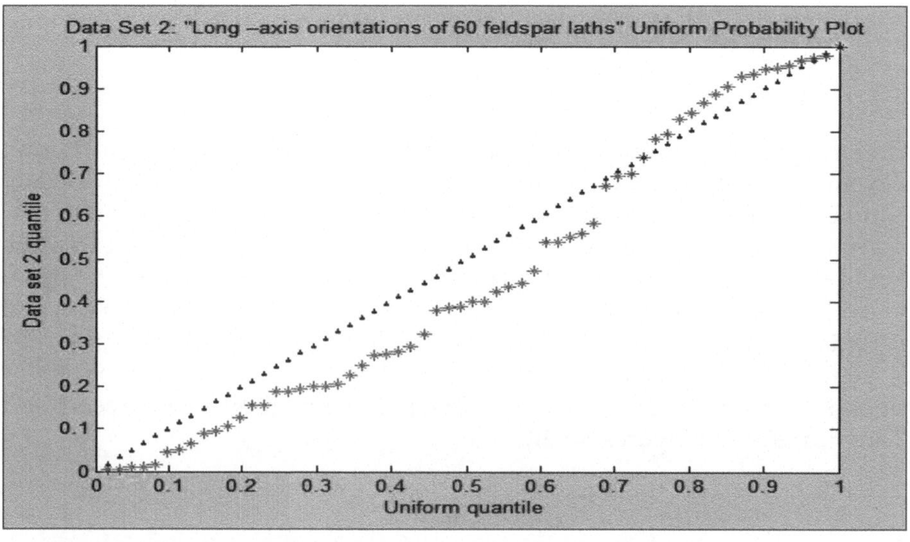

FIGURE 8.4
Uniform probability plot for data set 2.

8.3 Estimation of Parameters

The maximum likelihood method is used to estimate the parameters of the distributions under consideration for fitting the models (Subrahmanyam 2017). Likelihood equations of these two distributions for applying the Maximum Likelihood Estimation (MLE) method for estimating the parameters are as follows:

1. *Wrapped exponentiated inverted Weibull distribution*

   The probability density function of wrapped exponentiated inverted Weibull $g(\theta)$ is

$$g(\theta)=\sum_{k=0}^{\infty}\lambda c(\theta+2k\pi)^{-(c+1)}\left(e^{-(\theta+2k\pi)^{-c}}\right)^{\lambda} \quad (8.3.1)$$

   where $\theta \in (0,2\pi),\ c>0$, and $\lambda>0$

   The log likelihood function is

$$\begin{aligned}\log L&=\sum_{i=1}^{n}\log\left(\sum_{k=0}^{\infty}\lambda c(\theta_i+2k\pi)^{-(c+1)}\left(e^{-(\theta_i+2k\pi)^{-c}}\right)^{\lambda}\right)\\&=n\log\lambda+n\log c+\sum_{i=1}^{n}\log\sum_{k=0}^{\infty}\left[(\theta_i+2k\pi)^{-(c+1)}\left(e^{-(\theta_i+2k\pi)^{-c}}\right)^{\lambda}\right]\end{aligned} \quad (8.3.2)$$

   The likelihood equations are

$$\frac{\partial\log L}{\partial c}=\frac{n}{c}+\sum_{i=1}^{n}\left[\frac{\sum_{k=0}^{\infty}\left((\theta_i+2k\pi)^{-(c+1)}\left(e^{-(\theta_i+2k\pi)^{-c}}\right)^{\lambda}\log(\theta_i+2k\pi)\right)\left(\lambda(\theta_i+2k\pi)^{-c}-1\right)}{\sum_{k=0}^{\infty}\left[(\theta_i+2k\pi)^{-(c+1)}\left(e^{-(\theta_i+2k\pi)^{-c}}\right)^{\lambda}\right]}\right]=0 \quad (8.3.3)$$

$$\frac{\partial\log L}{\partial\lambda}=\frac{n}{\lambda}-\sum_{i=1}^{n}\left[\frac{\sum_{k=0}^{\infty}\left((\theta_i+2k\pi)^{-(2c+1)}\left(e^{-(\theta_i+2k\pi)^{-c}}\right)^{\lambda}\right)}{\sum_{k=0}^{\infty}\left((\theta_i+2k\pi)^{-(c+1)}\left(e^{-(\theta_i+2k\pi)^{-c}}\right)^{\lambda}\right)}\right]=0 \quad (8.3.4)$$

2. *Wrapped new Weibull–Pareto distribution*

   The probability density function of the wrapped new Weibull–Pareto distribution $g(\theta)$ is

$$g(\theta)=\sum_{k=0}^{\infty}\frac{c\delta}{\lambda}\left(\frac{\theta+2k\pi}{\lambda}\right)^{c-1}e^{-\delta\left(\frac{\theta+2k\pi}{\lambda}\right)^{c}} \tag{8.3.5}$$

   where $\theta\in(0,2\pi),\ c>0,\ \lambda>0$ and $\delta>0$.

   The log likelihood function is

$$\begin{aligned}\log L&=\sum_{i=1}^{n}\log\left(\sum_{k=0}^{\infty}\frac{c\delta}{\lambda}\left(\frac{\theta_i+2k\pi}{\lambda}\right)^{c-1}e^{-\delta\left(\frac{\theta_i+2k\pi}{\lambda}\right)^{c}}\right)\\&=n\log c+n\log\delta-n\log\lambda+\sum_{i=1}^{n}\log\left[\sum_{k=0}^{\infty}\left(\left(\frac{\theta_i+2k\pi}{\lambda}\right)^{c-1}\left(e^{-\delta\left(\frac{\theta_i+2k\pi}{\lambda}\right)^{c}}\right)\right)\right]\end{aligned} \tag{8.3.6}$$

   The likelihood equations are

$$\frac{\partial\log L}{\partial c}=\frac{n}{c}+\sum_{i=1}^{n}\left[\frac{\sum_{k=0}^{\infty}\left(\left(\frac{\theta_i+2k\pi}{\lambda}\right)^{c-1}\log\left(\frac{\theta_i+2k\pi}{\lambda}\right)e^{-\delta\left(\frac{\theta_i+2k\pi}{\lambda}\right)^{c}}\left[1-\delta\left(\frac{\theta_i+2k\pi}{\lambda}\right)^{c}\right]\right)}{\sum_{k=0}^{\infty}\left(\left(\frac{\theta_i+2k\pi}{\lambda}\right)^{c-1}e^{-\delta\left(\frac{\theta_i+2k\pi}{\lambda}\right)^{c}}\right)}\right]=0 \tag{8.3.7}$$

$$\frac{\partial\log L}{\partial\lambda}=-\frac{n}{\lambda}+\sum_{i=1}^{n}\left[\frac{\sum_{k=0}^{\infty}\left(\left(\frac{\theta_i+2k\pi}{\lambda}\right)^{c-1}e^{-\delta\left(\frac{\theta_i+2k\pi}{\lambda}\right)^{c}}\left[\frac{\delta c}{\lambda}\left(\frac{\theta_i+2k\pi}{\lambda}\right)^{c}+\left(\frac{1-c}{\lambda}\right)\right]\right)}{\sum_{k=0}^{\infty}\left(\left(\frac{\theta_i+2k\pi}{\lambda}\right)^{c-1}e^{-\delta\left(\frac{\theta_i+2k\pi}{\lambda}\right)^{c}}\right)}\right]=0 \tag{8.3.8}$$

$$\frac{\partial \log L}{\partial \delta}=\frac{n}{\delta}+\sum_{i=1}^{n}\left[\frac{\sum_{k=0}^{\infty}\left(\left(\frac{\theta_i+2k\pi}{\lambda}\right)^{2c-1}e^{-\delta\left(\frac{\theta_i+2k\pi}{\lambda}\right)^c}\right)}{\sum_{k=0}^{\infty}\left(\left(\frac{\theta_i+2k\pi}{\lambda}\right)^{c-1}e^{-\delta\left(\frac{\theta_i+2k\pi}{\lambda}\right)^c}\right)}\right]=0 \tag{8.3.9}$$

3. *Stereographic semicircular exponentiated inverted Weibull distribution*
   The probability density function of the stereographic semicircular exponentiated inverted Weibull distribution $g(\theta)$ is

$$g(\theta)=\frac{1}{2}\sec^2\left(\frac{\theta}{2}\right)\lambda c\left(\tan\left(\frac{\theta}{2}\right)\right)^{-(c+1)}\left(e^{-\left(\tan\left(\frac{\theta}{2}\right)\right)^{-c}}\right)^{\lambda} \tag{8.3.10}$$

where $\theta\in(0,\pi)$, $c>0$, and $\lambda>0$.

The log likelihood function is

$$\log L=\sum_{i=1}^{n}\log(g(\theta_i))$$

$$=\sum_{i=1}^{n}\log\left[\frac{1}{2}\sec^2\left(\frac{\theta_i}{2}\right)\lambda c\left(\tan\left(\frac{\theta_i}{2}\right)\right)^{-(c+1)}\left(e^{-\left(\tan\left(\frac{\theta_i}{2}\right)\right)^{-c}}\right)^{\lambda}\right]$$

$$=n\log\left(\frac{1}{2}\right)+2\sum_{i=1}^{n}\log\left(\sec\left(\frac{\theta_i}{2}\right)\right)+n\log\lambda+ \tag{8.3.11}$$

$$n\log c-(c+1)\sum_{i=1}^{n}\log\left(\tan\left(\frac{\theta_i}{2}\right)\right)-\lambda\sum_{i=1}^{n}\left(\tan\left(\frac{\theta_i}{2}\right)\right)^{-c}$$

The likelihood equations are

$$\frac{\partial \log L}{\partial c}=\frac{n}{c}-\sum_{i=1}^{n}\log\left(\tan\left(\frac{\theta_i}{2}\right)\right)+\lambda\sum_{i=1}^{n}\frac{\log\left(\tan\left(\frac{\theta_i}{2}\right)\right)}{\left(\tan\frac{\theta_i}{2}\right)^{c}}=0 \tag{8.3.12}$$

$$\frac{\partial \log L}{\partial \lambda}=\frac{n}{\lambda}-\sum_{i=1}^{n}\left(\tan\left(\frac{\theta_i}{2}\right)\right)^{-c}=0 \tag{8.3.13}$$

4. *Stereographic semicircular new Weibull–Pareto distribution*

The probability density function of stereographic semicircular new Weibull–Pareto distribution $g(\theta)$ is

$$g(\theta) = \frac{1}{2}\sec^2\left(\frac{\theta}{2}\right)\frac{c\delta}{\lambda}\left(\frac{\tan\left(\frac{\theta}{2}\right)}{\lambda}\right)^{c-1} e^{-\delta\left(\frac{\tan\left(\frac{\theta}{2}\right)}{\lambda}\right)^{c}} \tag{8.3.14}$$

where $\theta \in (0,\pi),\ c>0,\ \lambda>0, \text{and}\ \delta>0.$

The log likelihood function is

$$\log L = \sum_{i=1}^{n}\log\left[\frac{1}{2}\sec^2\left(\frac{\theta_i}{2}\right)\frac{c\delta}{\lambda}\left(\frac{\tan\left(\frac{\theta_i}{2}\right)}{\lambda}\right)^{c-1} e^{-\delta\left(\frac{\tan\left(\frac{\theta_i}{2}\right)}{\lambda}\right)^{c}}\right]$$

$$= n\log\left(\frac{1}{2}\right) + n\log c + n\log\delta - n\log\lambda$$

$$+2\sum_{i=1}^{n}\log\left(\sec\left(\frac{\theta_i}{2}\right)\right) + (c-1)\sum_{i=1}^{n}\log\left(\frac{\tan\left(\frac{\theta_i}{2}\right)}{\lambda}\right) - \delta\sum_{i=1}^{n}\left(\frac{\tan\left(\frac{\theta_i}{2}\right)}{\lambda}\right)^{c} \tag{8.3.15}$$

The likelihood equations are

$$\frac{\partial\log L}{\partial c} = \frac{n}{c} + \sum_{i=1}^{n}\log\left(\frac{\tan\left(\frac{\theta_i}{2}\right)}{\lambda}\right) - \delta\sum_{i=1}^{n}\left(\frac{\tan\left(\frac{\theta_i}{2}\right)}{\lambda}\right)^{c}\log\left(\frac{\tan\left(\frac{\theta_i}{2}\right)}{\lambda}\right) = 0 \tag{8.3.16}$$

$$\frac{\partial\log L}{\partial\delta} = \frac{n}{\delta} - \sum_{i=1}^{n}\left(\frac{\tan\left(\frac{\theta_i}{2}\right)}{\lambda}\right)^{c} = 0 \tag{8.3.17}$$

$$\frac{\partial\log L}{\partial\lambda} = \frac{\delta c}{\lambda^{c+1}}\sum_{i=1}^{n}\left(\tan\left(\frac{\theta_i}{2}\right)\right)^{c} - \frac{n}{\lambda} + \frac{n(1-c)}{\lambda} = 0 \tag{8.3.18}$$

The likelihood equations obtained are not analytically tractable. Hence, an optimization technique called *fmincon* available in MATLAB® is used for all the four distributions. A tool called *solver* of Microsoft Excel is used to find initial solutions to invoke the *fmincon*.

The initial solutions obtained using *solver* for the four distributions for the respective data sets are shown in Table 8.1.

The Maximum Likelihood (ML) estimates of the two circular models for the data set 1 and for two semicircular models for the data set 2 are obtained using *fmincon* are presented in Tables 8.2 and 8.3, respectively.

TABLE 8.1

Initial Values and Maximum of Log Likelihood Obtained through Solver

		Initial Values			Maximum of Log
S. No.	**Distribution**	c	λ	δ	**Likelihood**
1	Wrapped Exponentiated Inverted Weibull	1.0386	1.0585	—	−120.6282
2	Wrapped New Weibull–Pareto	1.2573	2.2637	1.0361	−118.7418
3	Stereographic Semicircular Expone ntiated Inverted Weibull	0.5140	0.5220		−72.0557
4	Stereographic Semicircular New Weibull–Pareto	0.5797	1.5429	0.9029	−67.1729

TABLE 8.2

ML Estimates for the Circular Models

		Parameters			Maximum of Log
S. No.	**Distribution**	c	λ	δ	**Likelihood**
1	Wrapped Exponentiated Inverted Weibull	1.0131	1.0773	—	−119.5338
2	Wrapped New Weibull–Pareto	1.2573	2.2000	0.9996	−118.7418

TABLE 8.3

ML Estimates for the Semicircular Models

		Parameters			Maximum of Log
S. No.	**Distribution**	c	λ	δ	**Likelihood**
1	Stereographic Semicircular Exponentiated Inverted Weibull	0.5140	0.5220	—	−72.0557
2	Stereographic Semicircular New Weibull–Pareto	0.5797	2.0848	1.0751	−67.1729

8.4 Goodness of Fit

Consideration of probability integral transformations on the line suggests the following analogue on the circle. Let G be the cumulative distribution function of a circular distribution, and suppose that an orientation and initial direction have been chosen. Then the probability integral transformation of the distribution is as the case may be the transformation of the circle/semi-circle, which takes θ to $2\pi G(\theta)/\pi G(\theta)$. If G is continuous, then the transformed random variable

$$U = 2\pi G(\theta) \bmod (2\pi) \tag{8.4.1}$$

is distributed uniformly on the circle. By means of the probability integral transformation (8.4.1), any test of uniformity on the circle gives rise to corresponding test of goodness of fit. For this purpose, the following tests for uniformity are adopted for verifying goodness of fit.

8.4.1 Kuiper's Test

A variety of tests for testing circular uniformity, on the lines of measuring the deviation between the observed and hypothesized cumulative distribution functions are available. The Kuiper test procedure extends this for the circular distributions.

Test Procedure: Given an angular data, first the observations are arranged in ascending order. Then these ordered observations $\theta_{(1)}, \theta_{(2)}, \ldots, \theta_{(n)}$ are augmented by $\theta_{(0)} = 0$ and $\theta_{(n+1)} = 2\pi$ in case of circular, and $\theta_{(0)} = 0$ and $\theta_{(n+1)} = \pi$ in case of semicircular models.

Then define $U_i = \frac{\theta_{(i)}}{2\pi}$ in case of circular and $U_i = \frac{\theta_{(i)}}{\pi}$, in case of semicircular, where $i = 1, 2, \ldots, n+1$.

$$D_n^+ = \max_{1 \le i \le n} \left\{ \frac{i}{n} - U_i \right\}$$

$$D_n^- = \max_{1 \le i \le n} \left\{ U_i - \frac{i-1}{n} \right\}$$

The Kuiper's statistics is

$$V_n = D_n^+ + D_n^- \tag{8.4.2}$$

It was modified by Stephens (1970) as

$$V_n^* = \sqrt{n}\ V_n \left(1 + \frac{0.155}{\sqrt{n}} + \frac{0.24}{n} \right) \tag{8.4.3}$$

TABLE 8.4

Upper Quantiles of V_n^*

α	0.1	0.05	0.025	0.01
V_n^*	1.62	1.747	1.862	2.001

Stephens (1970) showed that the distribution under uniformity of V_n^* varies very little with n, if $n \geq 8$. Some upper quantiles of V_n^* are given in Table 8.4, where α is the level of significance (LOS).

8.4.2 Watson's U^2 Test

Instead of taking the difference between the empirical distribution and the cumulative distribution of a uniform distribution by a variant of the maximum deviation, Watson (1961) considered (corrected) the mean square deviation. The test procedure of Watson's U^2 test is as follows:

Test Procedure: The first step is to compute

$$U^2 = \sum_{i=1}^{n}\left[U_i - \bar{U} - \frac{i-1/2}{n} + \frac{1}{2}\right]^2 + \frac{1}{12n}$$

where $\bar{U} = \dfrac{(U_1 + U_2 + \ldots + U_n)}{n}$

Stephens (1970) suggested a modified version of the statistics and is given by

$$U^{*2} = \left(U^2 - \frac{0.1}{n} + \frac{0.1}{n^2}\right)\left(1 + \frac{0.8}{n}\right) \tag{8.4.4}$$

Stephens (1970) showed that distribution under uniformity of U^{*2} varies very little with n for $n \geq 8$, and calculated upper quantiles of U^{*2} are given in Table 8.5, where α is the LOS.

8.4.2.1 Application of the Tests

The Kuiper's and Watson's U^2 tests (Mardia and Jupp 2000) are applied on all four angular models for testing goodness of fit.

TABLE 8.5

Upper Quantiles of U^{*2}

α	0.1	0.05	0.025	0.01
U^{*2}	0.152	0.187	0.221	0.267

TABLE 8.6

Goodness of Fit Tests for Circular Models (Data Set 1)

S. No.	Distribution	Kuiper's Test	Watson's U^2 Test
1	Wrapped Exponentiated Inverted Weibull	1.7133	0.1661
2	Wrapped New Weibull–Pareto	1.9481	0.2938

TABLE 8.7

Goodness of Fit Tests for Semicircular Models (Data Set 2)

S. No.	Distribution	Kuiper's Test	Watson's U^2 Test
1	Stereographic Semicircular Exponentiated Inverted Weibull	1.8917	0.2451
2	Stereographic Semicircular New Weibull–Pareto	1.5365	0.1783

Results of Kuiper's and Watson's U^2 tests are given in Tables 8.6 and 8.7 for circular and semicircular models, respectively.

Taking cut-off points of the respective test procedures from Devaraaj (2012) and comparing them with the results of Tables 8.6 and 8.7, it can be observed that

1. The wrapped exponentiated inverted Weibull distribution was found to be a good fit at 5% LOS as per both Kuiper's test and Watson's U^2 test for data set 1.
2. The wrapped new Weibull–Pareto was found to be a good fit at 1% LOS as per Kuiper's test for data set 1.
3. Stereographic semicircular exponentiated inverted Weibull was found to be a good fit at 1% LOS as per both Kuiper's test and Watson's U^2 test for data set 2.
4. Stereographic semicircular new Weibull–Pareto was found to be a good fit at 10% LOS as per Kuiper's test and at 5% LOS for Watson's U^2 test for data set 2.

Based on the observations, it can be seen that the circular and semicircular models under consideration were found to be good fits for the respective data sets, and in order to discriminate between good fit models, the procedures of maximum log likelihood (MLL), Akaike's information criteria (AIC), and Bayesian information criteria (BIC) are applied, and the results are presented in Tables 8.8 and 8.9.

TABLE 8.8
Results of Discriminating Procedures for Data Set 1

S. No.	Distribution	MLL	AIC	BIC
1	Wrapped Exponentiated Inverted Weibull	−119.53	243.07	247.73
2	Wrapped New Weibull–Pareto	−118.74	243.48	250.48

TABLE 8.9
Results of Discriminating Procedures for Data Set 2

S. No	Distribution	MLL	AIC	BIC
1	Stereographic Semicircular Exponentiated Inverted Weibull	−72.06	148.11	152.30
2	Stereographic Semicircular New Weibull–Pareto	−67.17	140.35	146.63

From Table 8.8, it can be noticed that the results obtained for the circular models are very close. However, based on Akaike Information Criterion (AIC) and Bayesian Information Criterion (BIC) criteria, it can be concluded that the wrapped exponentiated inverted Weibull distribution is a better model than the wrapped new Weibull–Pareto, for fitting to the data set 1.

From Table 8.9, for semicircular models, the stereographic semicircular new Weibull–Pareto distribution is clearly a better fit than the stereographic semicircular exponentiated inverted Weibull distribution based on all the three criteria, for fitting to the data set 2.

References

Devaraaj, V.J. 2012. Some Contributions to Circular Statistics, PhD dissertation, Acharya Nagarjuna University, Guntur, India.

Fisher, N.I. 1993. *Statistical Analysis of Circular Data*, Cambridge University Press, Cambridge, UK.

Jammalamadaka, S.R. and Sen Gupta, A. 2001. *Topics in Circular Statistics*, World Scientific Press, Singapore.

Mardia, K.V. and Jupp, P.E. 2000. *Directional Statistics*, John Wiley & Sons, Chichester, UK.

Stephens, M.A. 1970. Use of the Kolmogorov-Smirnov: Cramér-von Mises and Related Statistics Without Extensive Tables, *Journal of the Royal Statistical Society Series*, B, 32, 115–122.

Subrahmanyam, P.S. 2017. On Construction of Angular Models with Applications to Control Charts, PhD dissertation, Acharya Nagarjuna University, Guntur, India.

Watson, G.S. 1961. Goodness-of-fit tests on a circle, *Biometrika*, 48, 109–114.

9

Control Charts for Angular Data

9.1 Introduction

Quality is one of the most essential characteristics of any product. For manufacturers, quality is the most important marketing strategy, and for consumers it's the value for the money they spend on a product. Monitoring and controlling quality are a continuous process. Statistical process control (SPC) is a quality control technique using statistical methods that help in monitoring and controlling quality. Control charts are the most important tool used in SPC. Control charts are often used either to monitor an existing production process or to study a new one.

The application of the statistical methods in monitoring and controlling the quality of products dates back to the twentieth century. Walter Andrew Shewhart of the Bell Telephone Laboratories developed the concept of control charts.

A control chart is a graphical representation of the quality aspect of a process. The control charts are constructed using the sample data on the quality characteristic of a product, obtained from the production process. A control chart consists of a centre line (CL) representing the average of the quality characteristic. Another two lines horizontal to the CL, one above the CL and other one below the CL, are drawn on the control chart. The line that is above the CL is called upper control limit (UCL), and the one below the CL is called lower control limit (LCL). These control lines are constructed in such a way that if a process is under control, then most of the observations in the sample fall within the region covered between the UCL and LCL. If any point in the sample falls outside this region, then it is an indication that the process is out of control. An example of a control chart in the linear case is shown hereunder (Figure 9.1).

A process is said to be stable and consistent if the process is well within the predetermined quality norms. Control charts reveal whether the process in question is stable and consistent over a period of time. When a process is out of control, the reasons for such occurrence are explored, and corrective measures are initiated to bring the process back under control.

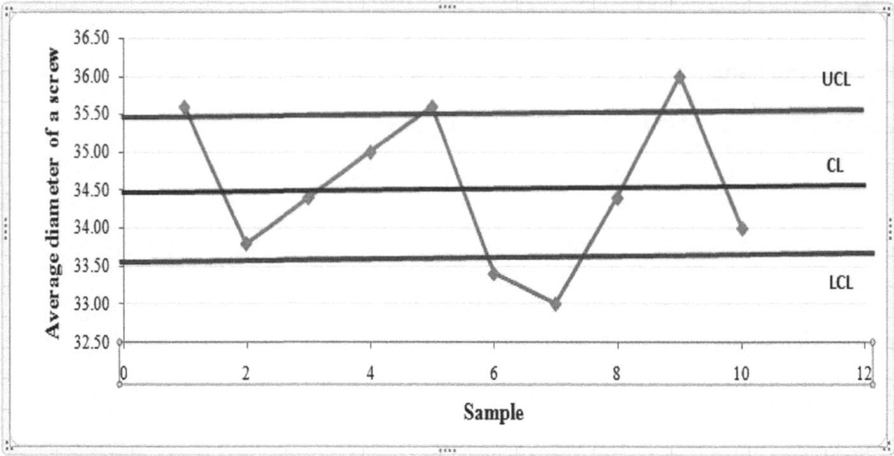

FIGURE 9.1
Linear control chart.

In statistical quality control, as detailed by Montgomery (2001), the concept of control charts is based on two kinds of variations: first kind of variations is due to chance causes or common causes, and the other variations are due to assignable causes or special causes. Chance cause variations are intrinsic, natural, stable and small variations in a system but very difficult to eliminate. Whereas the assignable cause variations are unusual, unexpected and large variations, but they can be eliminated easily, if diagnosed properly.

If a process has only chance cause variations in the system, then the process is said to be in control. If any assignable cause or special cause variations are also present, then the system is said to be out of control. Hence, it is important to separate the sources of variability into common causes and assignable causes.

Laha and Gupta (2011) applied this concept of control charts to angular distributions, such as von Mises, wrapped Cauchy, wrapped normal and Cardiod distributions and constructed central ray (CR), anticlockwise control ray (ACR) and clockwise control ray (CCR) to the above-mentioned circular distributions. They studied the impact of simulation size on the circular variances of ACR and CCR angles of these distributions, the effect of variation of parameters of circular distributions on the variances of ACR and CCR angles and the average run length (ARL) and median run length (MRL) of circular control charts.

The graph depicting the control charts for the mean direction in respect of circular distributions taken from Laha and Gupta (2011) is furnished in Figure 9.2.

Figure 9.2 represents the collection of concentric circles 1,2,...n, nth circle with radius n units. The CR represents the mean angle. The region bounded

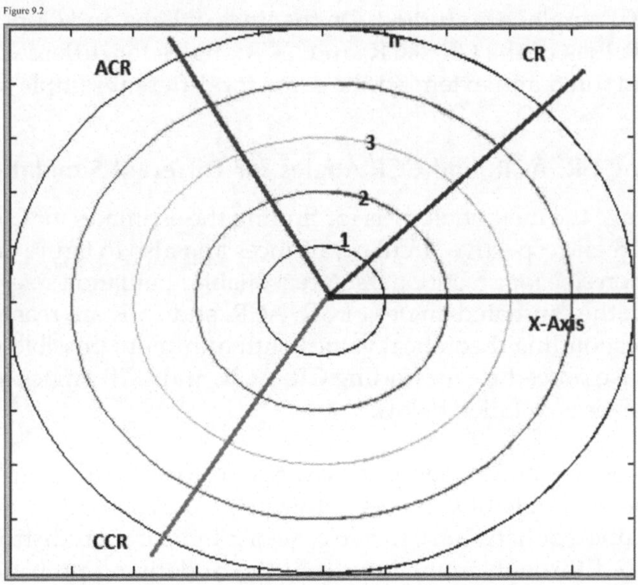

FIGURE 9.2
Circular control chart for mean direction.

by CCR–CR–ACR in the anticlockwise direction is the acceptance region, and the region bounded by ACR–CCR in the anticlockwise direction is the rejection region for the process for which the controls charts are constructed.

This chapter deals with extension of the concept of control charts to the circular distribution called wrapped exponentiated inverted Weibull distribution (WEIWD) and the stereographic semicircular new Weibull–Pareto distribution (SSCNWPD). The CR, ACR, and CCR are constructed for these distributions. Laha and Gupta (2011) studied the effects of simulation size on the circular variances of control rays. This idea is extended to sample size also, and the impact of both sample size and simulation size on the circular variances of ACR and CCR angles are studied. The procedure for computing circular mean and circular variance for a given set of angles is detailed in the Section 1.4.

9.2 Methodology Adopted for Construction of Control Charts

Random samples are drawn using the acceptance and rejection method from the circular and semicircular distributions for the construction of control charts (Subrahmanyam 2017), and the effects of sample size and simulation size on the control rays on the circular mean, circular variance of CR,

ACR, and CCR angles are studied. On the lines of Laha and Gupta (2011), the theoretical values of the CR, ACR, and CCR angles for different simulation sizes are computed and extended the same for different sample sizes.

9.2.1 Finding CR, ACR, and CCR Angles for Different Simulation Sizes

The objective of the methodology is for finding the estimates for CR, ACR, and CCR angles, their respective circular variances and also to figure out a reasonable simulation size for practitioners. A reasonable simulation size refers to the size at which the estimated angles of CR, ACR, and CCR are reasonably close to their corresponding theoretical values with minimum possible variance.

The stepwise procedure for finding CR, ACR, and CCR angles for different simulation sizes is detailed below.

1. For a selected simulation size, say, N and for a fixed sample size k, a random sample of k angles between 0 to 2π in case of a circular distribution or between 0 to π in case of a semicircular distribution is generated through simulation from the underlying probability distribution using the acceptance and rejection algorithm, as the distribution function is not in the closed form.
2. The circular mean μ_i of the ith subsample consisting of k observations is computed. Since N is the simulation size, Steps 1 and 2 are repeated for N times, and an array of circular means $\mu_1, \mu_2, \ldots, \mu_N$ is recorded.
3. The array of circular means is arranged in ascending order. The circular mean of all N circular means is computed, and the line joining the origin to the circular mean is the CR.
4. Taking the level of significance as 5% and by eliminating 0.025% circular means from both the ends of the sorted array of circular means, μ_U at the end of the array and μ_L at the beginning of the array are noted. Lines joining the origin to μ_U and μ_L are respectively ACR and CCR.

9.2.2 Finding Estimates and Variances of CR, ACR, and CCR Angles for a Given Simulation Size

After computing the CR, ACR and CCR angles for a selected simulation size, for finding estimates and circular variances for CR, ACR and CCR angles, the procedure detailed in Section 9.2.1 is iterated for a reasonable number of times (say 50). After the completion of 50 iterations, arrays of CR, ACR and CCR angles, each consisting of 50 angles, are obtained for a given simulation size.

In the second step, to get an estimated value of ACR angle for a given simulation size, the circular mean of the array of ACRs is computed. Similarly, the circular mean of the arrays of the CCR angle and CR angle give, respectively, estimated values of CCR and CR angles for a given simulation size. The circular variances of the arrays of CR, ACR, and CCR angles are computed.

It is pertinent to mention here that as the circular mean and the circular variance are simply the measure of mean direction and angular dispersion of a given set of angles, the terms "circular mean" and "circular variance" in this chapter for the mean and variance of angles of semicircular distributions are also for the sake of convenience.

Changing the simulation size *N* from 250, 500, 1000, 2000, and up to 10,000, the corresponding estimated values for CR, ACR, and CCR angles and their circular variances are computed.

This procedure is applied on the two selected distributions and again for a distribution for different cases each with different set of values to the parameters.

9.2.3 Finding CR, ACR, and CCR Angles for Different Sample Sizes

As already stated, the procedures detailed in previous sections are extended for different sample sizes. The objective of the methodology is for finding the estimates for CR, ACR, and CCR angles, their respective circular variances and also to figure out a reasonable sample size for practitioners. A reasonable sample size refers to the sample size at which the estimated angles of ACR and CCR are reasonably close to their CR angle with minimum possible variance.

For finding CR, ACR, and CCR angles for a given sample size, the procedure detailed in 9.2.1, by fixing the simulation size to 1000 and varying the sample size from 5, 10, and up to 100 in steps of 10, is applied.

9.2.4 Finding Estimates and Variances of CR, ACR, and CCR Angles for a Given Sample Size

After computing the CR, ACR, and CCR angles for a selected sample size, for finding estimates and circular variances of CR, ACR, and CCR angles, the procedure detailed in Section 9.2.3 is iterated 50 times. After completion of 50 iterations, arrays of CR, ACR, and CCR angles, each consisting of 50 angles for a given sample size, are obtained.

Following the procedure similar to that of Section 9.2.2, this time by fixing the simulation size at 1000 and changing the sample size *k* from 5, 10, and up to 100 in steps of 10, sample sizewise, estimates and circular variances of CR, ACR and CCR angles are computed. The same procedure on selected distributions and again for a distribution for different cases, each with different set of values to the parameters, is applied.

9.2.5 Finding Theoretical Values of CR, ACR, and CCR Angles

The theoretical value of the CR, ACR, and CCR angles for the distributions are also computed by taking a sufficiently large value for *N*, the simulation size. The theoretical values for CR, ACR, and CCR angles by taking simulation size as ***one million*** are computed and tabulated.

9.3 Construction of Control Charts for Circular Distributions

In this section, the control charts for the circular distribution called WEIWD at different values of parameters are constructed. The theoretical values for CR, ACR, and CCR angles are obtained. The acceptance region and rejection region at the theoretical values are identified, and the expansion and contraction of the acceptance region with respect to change in parameter values are studied. The estimates for CR, ACR, and CCR angles and their respective circular variances are computed. The impact of simulation size and sample size on the ACR, and CCR angles and on their circular variances are also studied. An effort to figure out a reasonable simulation size as well as sample size for finding estimate values for CR, ACR, and CCR angles in case of circular distribution is also made.

In this section, for finding CR, ACR, and CCR angles for different simulation sizes, the sample size is fixed at 5. For finding CR, ACR, and CCR angles for different sample sizes, the simulation size is fixed at 1000. As this section deals with circular distributions, random samples consisting of angles between 0 and 2π are generated through simulation from the underlying probability distribution.

All angles are given in degrees. To show the acceptance and rejection region for a distribution graphically, east as zero direction and anticlockwise as sense of rotation are adopted. The notations for circular variance of central ray (CVCR), circular variance of anticlockwise control ray (CVACR) and circular variance of clockwise control ray (CVCCR) are used in the column headings of the tables.

9.3.1 Control Charts for Wrapped Exponentiated Inverted Weibull Distribution

The probability distribution function of WEIWD is

$$g(\theta) = \sum_{k=0}^{\infty} \lambda c(\theta+2\pi k)^{-(c+1)}\left(e^{-(\theta+2\pi k)^{-c}}\right)^{\lambda}$$

where $\theta \in (0,2\pi), c > 0$, and $\lambda > 0$.

1. **Theoretical values obtained for the CR, ACR, and CCR angles for WEIWD for different cases.**

   Using the procedure explained in Section 9.2.5, the theoretical values for various cases are computed and tabulated in Table 9.1. The corresponding graphs, casewise, are presented.

   The acceptance region in the Figure 9.3 is the region bounded by CCR–CR–ACR, i.e. from 50.260 to 169.920 in the anticlockwise

TABLE 9.1
Theoretical Values for Various Cases

Case	c	λ	CR	ACR	CCR
1	2	2	96.86	169.92	50.26
2	2	3	113.32	205.44	56.29
3	3	2	87.65	126.91	60.53
4	3	3	98.88	142.93	68.17
5	4	3	88.61	118.57	67.83

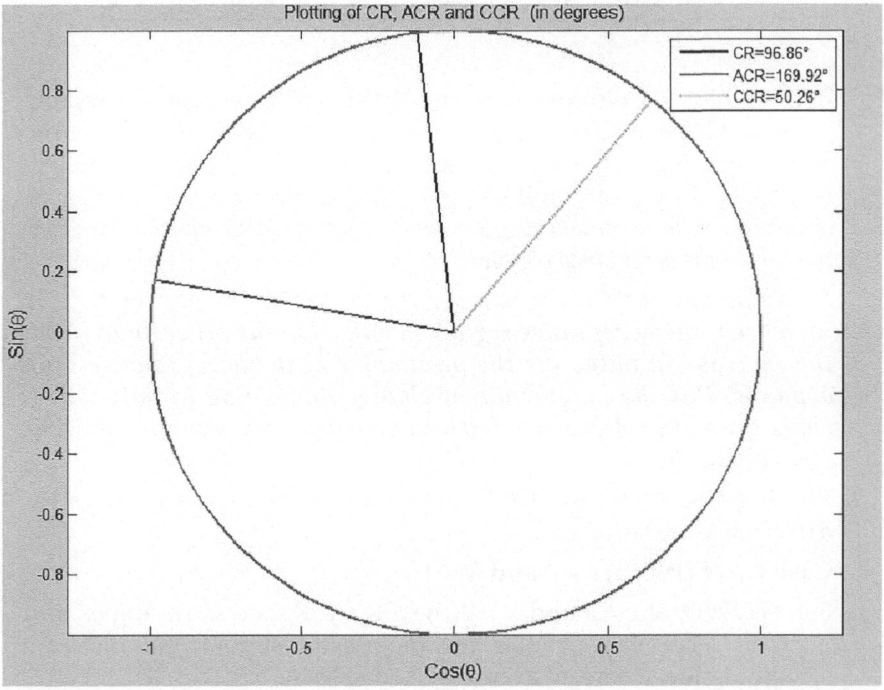

FIGURE 9.3
Graph of CR, ACR, and CCR for Case 1.

direction taking east as zero direction. Similar graphs can be drawn for other cases, and the same interpretation provided for Figure 9.3 holds good for them also.

2. **Analysis of theoretical values obtained for CR, ACR, and CCR angles.**

Theoretical values for CR, ACR, and CCR angles for the WEIWD at different values for parameters c and λ are computed to study the expansion and contraction of the acceptance region in response to the changes in the value of parameters. The results are furnished in Table 9.2.

TABLE 9.2

Theoretical Values and Acceptance Region of WEIWD

S. No.	c	λ	CR	ACR	CCR	ACR–CCR	Change
(1)	(2)	(3)	(4)	(5)	(6)	(7) = (5)–(6)	(8)
1	2	2	96.86	169.92	50.26	119.66	—
2	2	3	113.32	205.44	56.29	149.15	29.49
3	3	3	98.88	142.93	68.17	74.76	–74.39
4	3	4	107.58	155.55	74.02	81.52	6.77
5	4	4	94.85	126.69	72.69	54.00	–27.53
6	4	5	99.93	133.16	76.56	56.60	2.60

In the Table 9.2, Column 7 indicates the size of the acceptance region, whereas Column 8 indicates the change occurred in the size of the acceptance region when the value of one of the two parameters is increased by a one unit. *A positive value in Column 8 indicates the expansion of the acceptance region, and a negative value indicates the contraction of the acceptance region.*

It can be noticed that whenever the value for the parameter c is increased, the acceptance region shrinks. On the other hand, with the increase in value for the parameter λ, the acceptance region expands. But the magnitude of change happening in both directions gradually decreases with increasing values for the parameters c and λ.

3. **Finding CR, ACR, and CCR angles and respective variances for different simulation sizes.**

   Case 1: WEIWD at $c = 2$ and $\lambda = 2$

   For WEIWD, at $c = 2$ and $\lambda = 2$ the CR, ACR, and CCR angles and their respective circular variances are computed for different simulation sizes and are furnished in Table 9.3.

   The graph of circular variances for ACR angles for different simulation sizes is obtained as in Figure 9.4.

   The graph of circular variances for CCR angles for different simulation sizes is obtained as in Figure 9.5.

   Comparing the theoretical values of CR, ACR, and CCR angles for WEIWD, at $c = 2$ and $\lambda = 2$ tabulated at Table 9.2, with values obtained for CR, ACR, and CCR angles for different simulation sizes at Table 9.3, it can be observed that with the increasing simulation size, the CR, ACR, and CCR angles are close to their respective theoretical values.

   Also from Figures 9.4 and 9.5, the circular variances of ACR and CCR are moderate when the simulation size is around 250 to 500,

TABLE 9.3
CR, ACR and CCR Angles with Circular Variance for Case 1

Simulation Size	CR	CVCR	ACR	CVACR	CCR	CVCCR
250	96.99	0.02	173.51	1.55	51.08	0.17
500	96.96	0.01	170.15	0.65	51.69	0.10
1000	96.80	0.01	168.94	0.28	50.52	0.05
2000	96.98	0.00	170.03	0.20	50.47	0.03
3000	96.90	0.00	170.02	0.08	50.31	0.02
4000	96.79	0.00	169.82	0.10	50.37	0.01
5000	96.83	0.00	169.98	0.08	50.22	0.01
6000	96.86	0.00	169.31	0.04	50.20	0.01
7000	96.82	0.00	169.52	0.04	50.01	0.01
8000	96.92	0.00	169.76	0.03	50.52	0.01
9000	96.89	0.00	170.18	0.04	50.31	0.01
10000	96.88	0.00	169.97	0.04	50.19	0.01

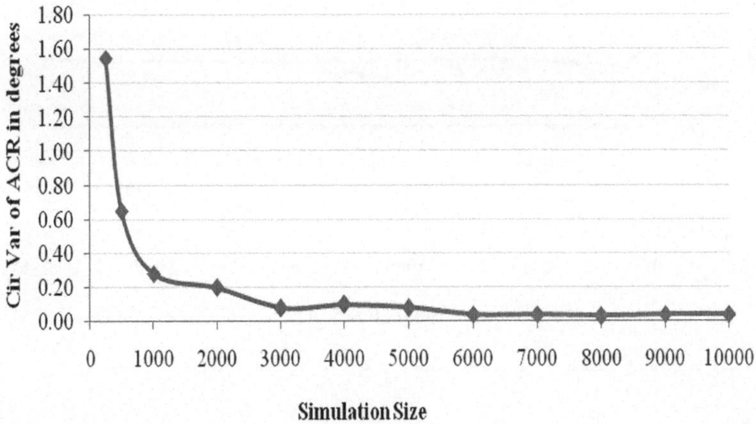

FIGURE 9.4
Circular variance of ACR angles for Case 1.

but with the increasing simulation size, especially after 1000, a sharp reduction in the circular variances can be seen and finally reaching to zero for higher simulation sizes.

A similar trend is observed for Cases 2, 3, 4, and 5 though not included here.

Casewise ACR, and CCR angles for different simulation sizes

The ACR angles and CCR angles with different simulation sizes are depicted in graphs and are furnished in Figures 9.6 and 9.7.

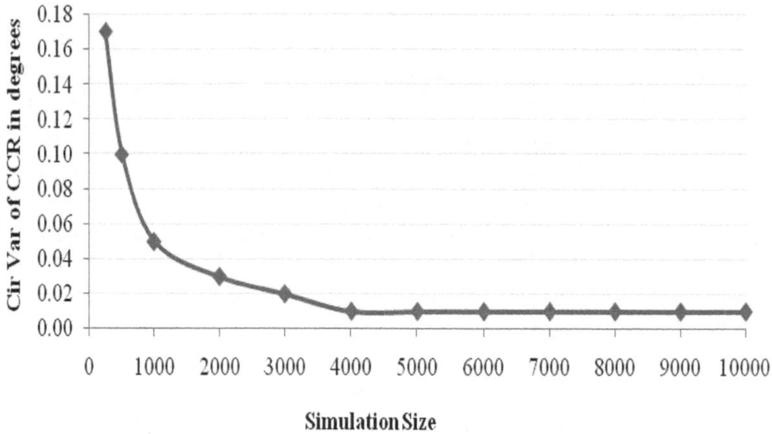

FIGURE 9.5
Circular variance of CCR angles for Case 1.

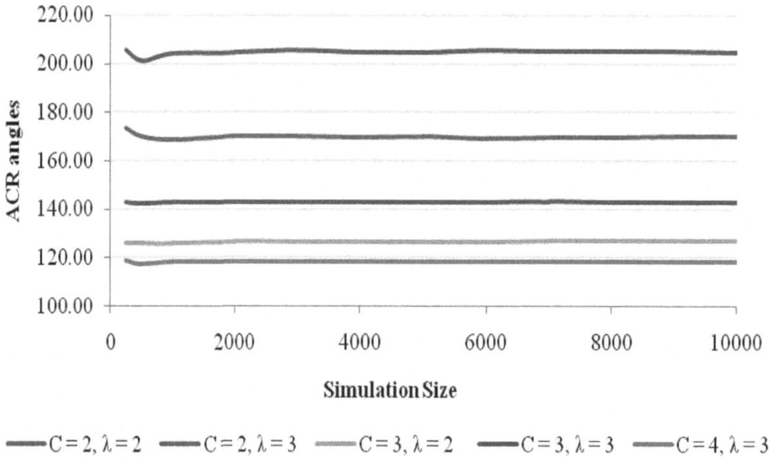

FIGURE 9.6
ACR angles of WEIWD for different simulation sizes.

4. Finding CR, ACR, and CCR angles for different sample sizes

Case 1: WEIWD at $c = 2$ and $\lambda = 2$

For WEIWD at $c = 2$ and $\lambda = 2$ the CR, ACR, and CCR for different sample sizes are furnished in Table 9.4.

The graph of CR, ACR, and CCR angle for different sample sizes is furnished in Figure 9.8.

The graph of circular variances for ACR angles for different sample sizes is obtained as in Figure 9.9.

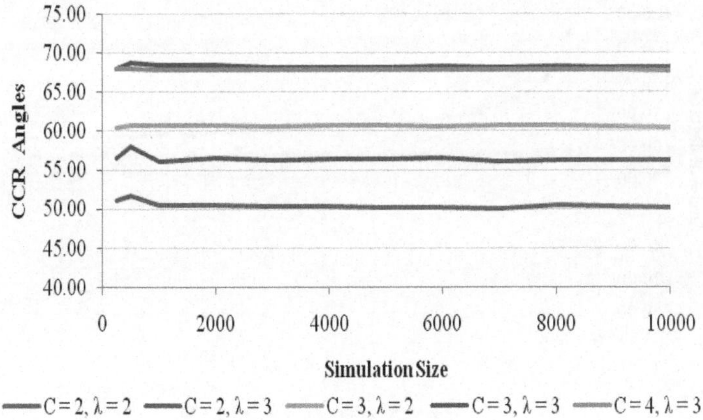

FIGURE 9.7
CCR angles of WEIWD for different simulation sizes.

TABLE 9.4
CR, ACR, and CCR Angles with Circular Variances for Case 1

Sample Size	CR	CVCR	ACR	CVACR	CCR	CVCCR
5	96.70	0.01	170.46	0.31	50.81	0.06
10	96.34	0.00	138.72	0.07	64.04	0.01
20	95.75	0.00	122.81	0.02	73.25	0.01
30	95.56	0.00	116.98	0.01	77.17	0.00
40	95.41	0.00	113.71	0.00	79.24	0.00
50	95.31	0.00	111.54	0.00	80.98	0.00
60	95.36	0.00	109.99	0.00	82.29	0.00
70	95.33	0.00	108.91	0.00	83.15	0.00
80	95.28	0.00	107.87	0.00	83.75	0.00
90	95.25	0.00	106.96	0.00	84.50	0.00
100	95.30	0.00	106.45	0.00	85.11	0.00

The graph of circular variances for CCR angles for different sample sizes is obtained as in Figure 9.10.

From Figure 9.8, it is clear that with increasing sample size the ACR and CCR angles are coming close to the CR angle. Also from Figures 9.9 and 9.10, the circular variances of ACR and CCR are reaching zero with increasing sample size after a sharp reduction in circular variance at sample size 10.

A similar trend is observed for Cases 2, 3, 4, and 5 though not included here.

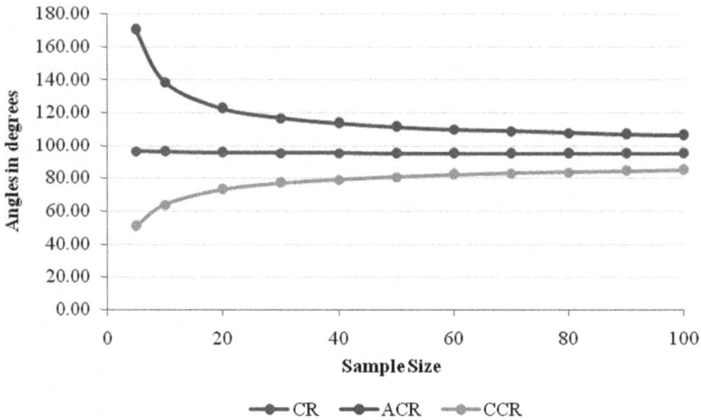

FIGURE 9.8
CR, ACR, and CCR angles for Case 1.

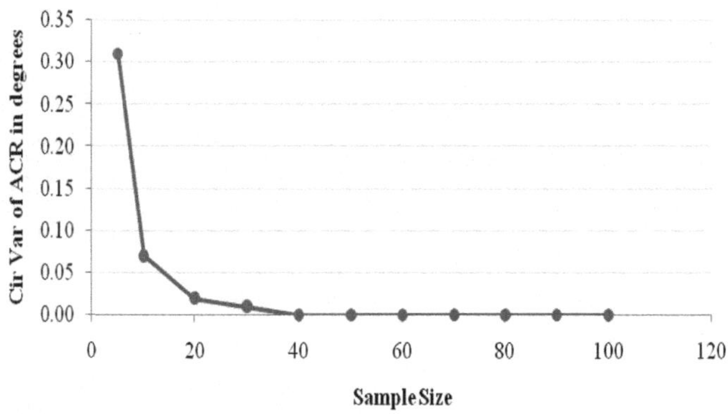

FIGURE 9.9
Circular variance of ACR angles for Case 1.

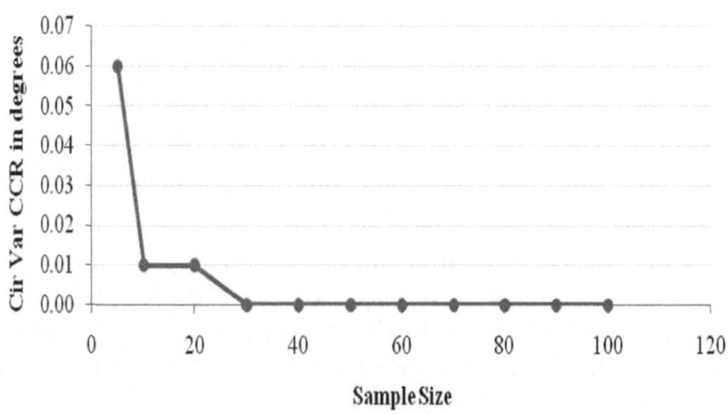

FIGURE 9.10
Circular variance of CCR angles for Case 1.

9.4 Construction of Control Charts for Semicircular Distributions

In this section, the control charts for the stereographic semicircular new Weibull–Pareto distribution at different sets of values of parameters are constructed. The theoretical values for CR, ACR, and CCR angles are obtained. The acceptance region and rejection region at the theoretical values are identified, and the expansion and contraction of the acceptance region with respect to changes in parameter values are studied. The estimates for ACR and CCR angles and their respective circular variances are computed for each one of these distributions and for different sets of parameters. The impact of simulation size and sample size on the ACR and CCR angles and on their circular variance is also studied. An effort to figure out a reasonable simulation size as well as sample size for finding estimate values for CR, ACR, and CCR angles in case of semicircular distributions is also made.

In this section, for all the computations, the level of significance α as 5% and number of iterations as 50 are considered. For finding CR, ACR, and CCR angles for different simulation sizes, the sample size is fixed at 5. For finding CR, ACR, and CCR angles for different sample sizes, the simulation size is fixed at 1000.

As this section is dealing with semicircular distributions, random samples consisting of angles between 0 and π are generated through simulation from the underlying probability distribution.

9.4.1 Control Charts for Stereographic Semicircular New Weibull–Pareto Distribution

The probability distribution function of the SSCNWPD is

$$g(\theta)=\frac{1}{2}\mathrm{Sec}^2\left(\frac{\theta}{2}\right)\frac{c\,\delta}{\lambda}\left(\frac{\tan\left(\frac{\theta}{2}\right)}{\lambda}\right)^{c-1}e^{-\delta\left(\frac{\tan\left(\frac{\theta}{2}\right)}{\lambda}\right)^{c}}$$

where $\theta\in(0,\pi)$, $c>0$, $\lambda>0$, and $\delta>0$.

1. **Theoretical values obtained for CR, ACR, and CCR angles for SSCNWPD for different cases.**

   Using the procedure explained in Section 9.2.5, the theoretical values for various cases were computed and tabulated in Table 9.5. The corresponding graphs, casewise, are presented.

TABLE 9.5

Theoretical Values for Various Cases

Case	c	δ	λ	CR	ACR	CCR
1	2	2	2	94.69	119.08	65.31
2	2	3	2	84.44	109.01	56.50
3	3	2	2	105.03	121.86	83.64
4	3	3	2	97.94	115.19	76.67
5	3	3	3	118.41	133.82	97.65

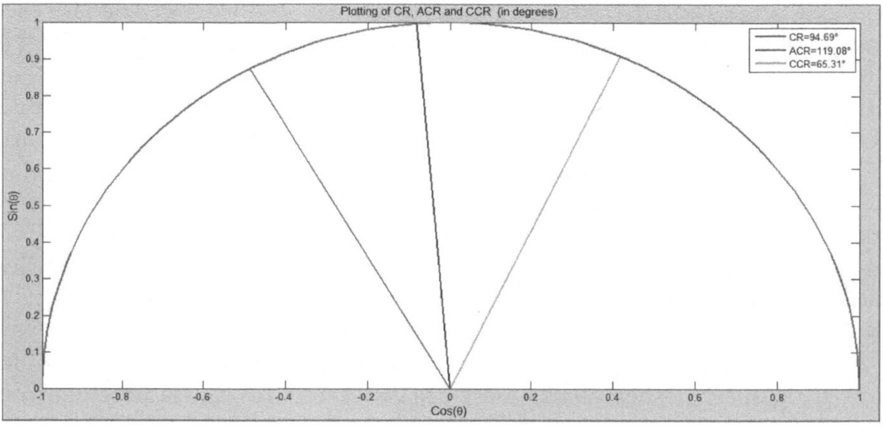

FIGURE 9.11
Graph of CR, ACR and CCR for Case 1.

From the Figure 9.11, the acceptance region is from 94.690 to 119.080 in the anticlockwise direction with east as the zero direction. The region excluding this in the semicircle is the rejection region.

2. **Analysis of theoretical values obtained for CR, ACR, and CCR angles.**

Theoretical values for CR, ACR, and CCR angles for the SSCNWPD at different values for parameters c,δ, and λ are computed to study the expansion and contraction of the acceptance region in response to the changes in the value of parameters. The results are furnished in Table 9.6.

In Table 9.6, Column 8 indicates the size of the acceptance region, whereas Column 9 indicates the change occurred in the size of the acceptance region when the value of one of the three parameters is increased by a one unit. *A positive value in Column 9 indicates the expansion of the acceptance region, and a negative value indicates contraction of the acceptance region.*

It can be observed from the Table 9.6 that with increasing the value of parameter c the acceptance region shrinks. Increasing the value

TABLE 9.6

Theoretical and Acceptance Region for SSCNWPD

S. No.	c	δ	λ	CR	ACR	CCR	ACR–CCR	Change
(1)	(2)	(3)	(4)	(5)	(6)	(7)	(8) = (6)–(7)	(9)
1	2	2	2	94.69	119.08	65.31	53.76	—
2	2	3	2	84.44	109.01	56.50	52.51	−1.26
3	2	3	3	104.77	128.30	74.77	53.54	1.03
4	3	3	3	118.41	133.82	97.65	36.17	−17.37
5	3	4	3	113.76	129.76	92.62	37.15	0.98
6	3	4	4	127.02	141.14	107.26	33.88	−3.27
7	4	4	4	134.38	144.22	120.23	23.98	−9.90
8	4	5	4	132.13	142.33	117.67	24.66	0.68
9	4	5	5	140.64	149.43	127.62	21.81	−2.85
10	5	5	5	144.75	151.25	135.07	16.18	−5.62
11	5	6	5	143.55	150.24	133.63	16.61	0.42
12	5	6	6	149.23	155.01	140.43	14.59	−2.02

of parameter λ also moderately decreases the acceptance region except when the value of λ increases from 2 to 3. Whereas increasing the value of parameter δ moderately increases the acceptance region except when the value of δ increases from 2 to 3.

3. **Finding CR, ACR, and CCR angles for different simulation sizes.**

Case 1: SSCNWPD at $C = 2$, $\delta = 2$, and $\lambda = 2$

For SSCNWPD at $C = 2$, $\delta = 2$, and $\lambda = 2$ the CR, ACR, and CCR angles are computed for different simulation sizes and are furnished in Table 9.7.

TABLE 9.7

CR, ACR, and CCR Angles with Circular Variance for Case 1

Simulation Size	CR	CVCR	ACR	CVACR	CCR	CVCCR
250	94.70	0.01	118.63	0.02	66.29	0.07
500	94.82	0.00	118.99	0.01	66.17	0.02
1000	94.75	0.00	118.89	0.01	65.89	0.01
2000	94.73	0.00	119.08	0.00	65.19	0.01
3000	94.68	0.00	119.12	0.00	65.43	0.01
4000	94.72	0.00	119.04	0.00	65.53	0.00
5000	94.71	0.00	119.15	0.00	65.39	0.00
6000	94.69	0.00	119.14	0.00	65.55	0.00
7000	94.69	0.00	119.12	0.00	65.39	0.00
8000	94.70	0.00	119.09	0.00	65.39	0.00
9000	94.67	0.00	119.02	0.00	65.29	0.00
10000	94.69	0.00	119.02	0.00	65.41	0.00

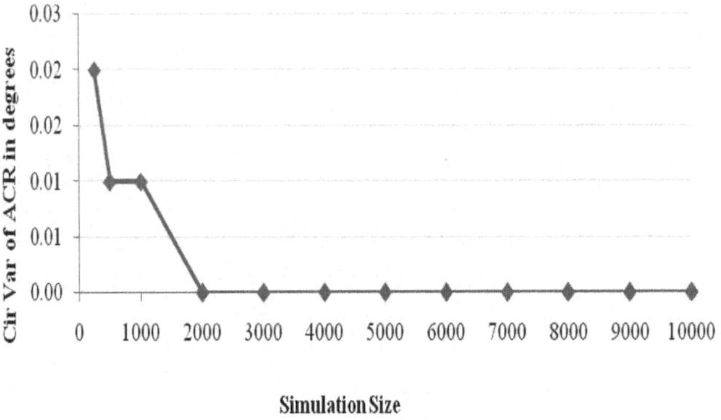

FIGURE 9.12
Circular variance of ACR angles for Case 1.

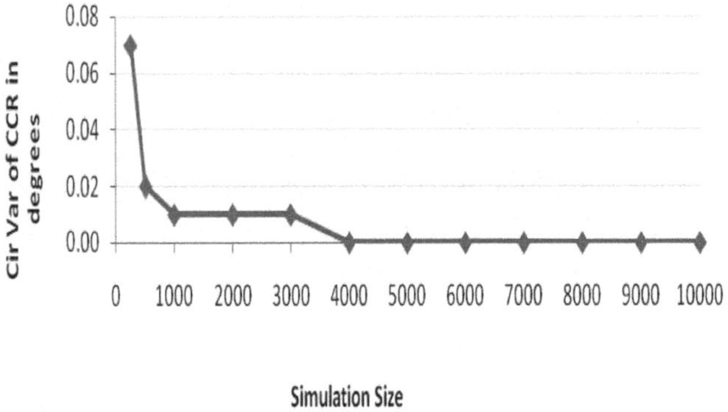

FIGURE 9.13
Circular variance of CCR angles for Case 1.

The graph of circular variances for ACR angles for different simulation sizes is obtained as in Figure 9.12.

The graph of circular variances for CCR angles for different simulation sizes is obtained as in Figure 9.13.

Comparing the theoretical values tabulated at Table 9.6 ***for CR, ACR and CCR for SSCNWPD at*** $C = 2$, $\delta = 2$, *and* $\lambda = 2$ ***with values obtained for CR, ACR, and CCR for different simulation sizes furnished above at Table 9.7, it can be observed that with increasing simulation size, the CR, ACR, and CCR angles are close to their respective theoretical values. Also from Figures 9.12 and 9.13, the circular variances of ACR and CCR are more when the simulation size is small, but with the simulation size exceeding 1000, it can be noticed with a sharp reduction the circular variances reaching zero for both ACR and CCR angles.***

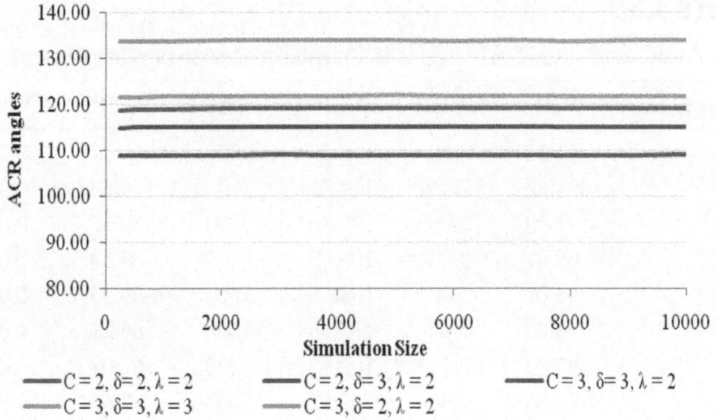

FIGURE 9.14
ACR angles of SSCNWPD for different simulation sizes.

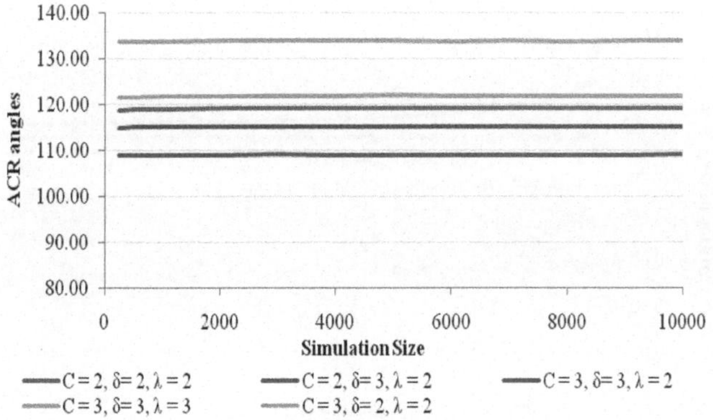

FIGURE 9.15
CCR angles of SSCNWPD for different simulation sizes.

A similar trend is observed for Cases 2, 3, 4, and 5 though not included here.

Casewise ACR and CCR angles for different simulation sizes

The ACR angles and CCR angles with different simulation sizes are depicted in graphs and are furnished in Figures 9.14 and 9.15.

4. **Finding CR, ACR, and CCR angles for different sample sizes.**

Case 1: SSCNWPD at $C = 2$, $\delta = 2$, and $\lambda = 2$

For SSCNWPD at $C = 2$, $\delta = 2$, and $\lambda = 2$ the CR, ACR, and CCR for different sample sizes are furnished in Table 9.8.

The graph of CR, ACR, and CCR angle for different sample sizes is furnished in Figure 9.16.

The graph of circular variances for CCR angles for different sample sizes is obtained as in Figure 9.17.

TABLE 9.8

CR, ACR, and CC R Angles with Circular Variance for Case 1

Sample Size	CR	CVCR	ACR	CVACR	CCR	CVCCR
5	94.62	0.00	118.85	0.00	65.51	0.01
10	94.90	0.00	112.72	0.00	74.63	0.01
20	94.98	0.00	107.80	0.00	80.87	0.00
30	95.01	0.00	105.65	0.00	83.59	0.00
40	95.01	0.00	104.25	0.00	85.15	0.00
50	95.09	0.00	103.35	0.00	86.26	0.00
60	95.04	0.00	102.65	0.00	87.04	0.00
70	95.05	0.00	102.12	0.00	87.65	0.00
80	95.05	0.00	101.69	0.00	88.16	0.00
90	95.09	0.00	101.33	0.00	88.56	0.00
100	95.10	0.00	101.06	0.00	88.91	0.00

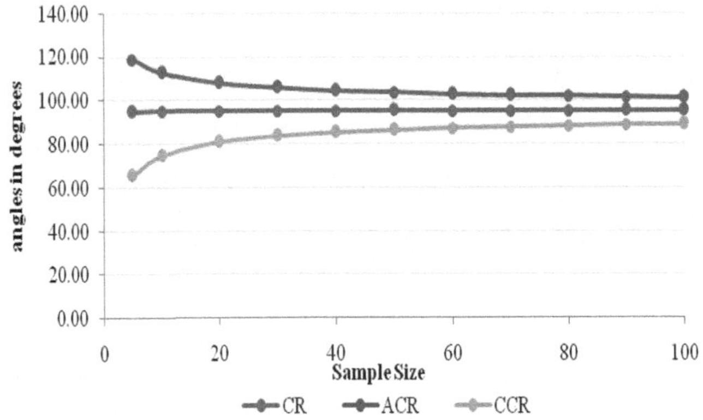

FIGURE 9.16
CR, ACR, and CCR angles for Case 1.

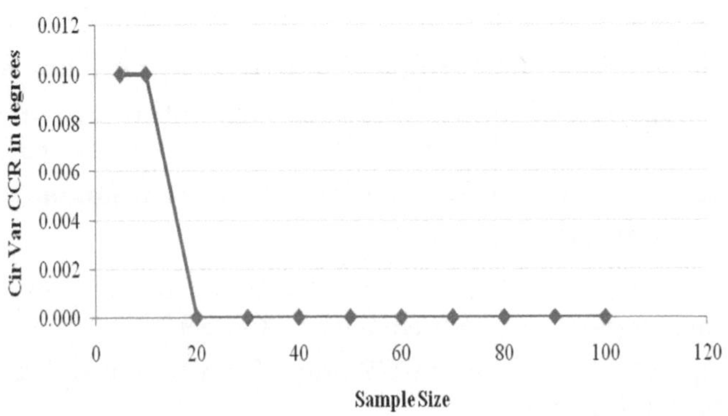

FIGURE 9.17
Circular variance of CCR angles for Case 1.

From Figure 9.16, it can be observed that with increasing sample size, the ACR and CCR angles are coming close to CR angle. Also from Figure 9.17, the circular variance of CCR angles is moderate when the sample size is small, but with the increasing sample size, especially after 10, the circular variances CCR angles become zero. The circular variances of ACR angles are zero for all sample sizes in this case.

A similar trend is observed for Cases 2, 3, 4, and 5 though not included here.

References

Laha, A.K. and Gupta A., 2011. Statistical Quality Control of Directional Data, *2nd IIMA International Conference on Advanced Data Analysis, Business analytics and Intelligence*, Ahmedabad, India, 8–9 January 2011.

Montgomery, D.C. 2001. *Introduction to Statistical Quality Control*, 3rd ed. John Wiley & Sons, Singapore.

Subrahmanyam, P.S. 2017. On Construction of Angular Models with Applications to Control Charts, PhD dissertation, Acharya Nagarjuna University, Guntur, India.

Appendix

A1.1 MATLAB Programme Listings

Notations Used in the Programs

th - θ
lambda - λ
pi - π
phi - φ - characteristic function
real(phi) – real (φ)
imag(phi) – imaginary (φ)
alpha , beta - α_1, β_1 - first order trigonometric moments
rho - ρ_1 resultant length
mu - μ_o - mean direction
alphag(p), betag(p) - α_p^* , β_p^* central trigonometric moments
v0 – V_0 - circular variance
sig - σ_o - standard deviation
gamma1 – γ_1^o - skewness
gamma2 - γ_2^o - kurtosis

The following are the typical programs given for plotting graphs of pdf, cdf, and characteristic function, evaluating population characteristics, goodness of fit, and control charts of Wrapped Exponentiated Inverted Weibull Distribution and Stereographic Semicircular Exponentiated Inverted Weibull Distribution. With suitable modifications, they can me adopted for other models also.

1. Program for the graph of pdf (linear representation) of wrapped exponentiated inverted Weibull distribution:

```
clc
th=0.000001:0.001:2*pi;
c=1;
lambda=2.0;
n=3;
g=0;
for k=0:n
    g=g+((lambda.*c).*((th+2*k*pi)).^(-(c+1))).
    *(exp(-(th+2*k*pi). ^(-c)).^lambda);
end
```

```
plot(th,g,'k')
hold on
c=1.5;
lambda=2.0;
g=0;
for k=0:n
    g=g+((lambda.*c).*((th+2*k*pi)).^(-(c+1))).
    *(exp(-(th+2*k*pi).^ (-c)).^lambda);
end
plot(th,g,'c')
hold on
c=2.0;
lambda=2.0;
g=0;
for k=0:n
    g=g+((lambda.*c).*((th+2*k*pi)).^(-(c+1))).
    *(exp(-(th+2*k*pi).^(-c)).^lambda);
end
plot(th,g,'b')
hold on
c=2.5;
lambda=2.0;
g=0;
for k=0:n
    g=g+((lambda.*c).*((th+2*k*pi)).^(-(c+1))).
    *(exp(-(th+2*k*pi).^(-c)).^lambda);
end
plot(th,g,'m')
hold off
xlabel('\theta')
ylabel('g(\theta)')
title('PDF of Wrapped Exponentiated Inverted Weibull
Distribution(Linear Representation)')
legend('C=1.0,\lambda=2.0','C=1.5,\lambda=2.0','C=2.0,\
lambda=2.0','C=2.5,\lambda=2.0')
```

2. Program for the graph of pdf (circular representation) of wrapped new Weibull–Pareto distribution:

```
th=0.001:0.001:2*pi;
c=2.0;
lambda=2.0;
del=3.0;
n=3;
g=0;
for k=0:n
    g=g+(((c.*del)./lambda).*(((th+2*k*pi)/lambda).^(c-1)).
    *(exp((-del).*((th+2*k*pi)/lambda).^c)));
end
```

```
r=1+g;
x=r.*cos(th);
y=r.*sin(th);
plot(x,y,'m')
hold on
c=2.5;
lambda=2.0;
del=2.5;
g=0;
for k=0:n
   g=g+(((c.*del)./lambda).*(((th+2*k*pi)/lambda).^(c-1)).
   *(exp((-del).*((th+2*k*pi)/lambda).^c)));
end
r=1+g;
x=r.*cos(th);
y=r.*sin(th);
plot(x,y,'b')
hold on
c=3.0;
lambda=2.0;
del=2.0;
g=0;
for k=0:n
   g=g+(((c.*del)./lambda).*(((th+2*k*pi)/lambda).^(c-1)).
   *(exp((-del).*((th+2*k*pi)/lambda).^c)));
end
r=1+g;
x=r.*cos(th);
y=r.*sin(th);
plot(x,y,'c')
hold on
c=3.5;
lambda=2.0;
del=1.5;
g=0;
for k=0:n
   g=g+(((c.*del)./lambda).*(((th+2*k*pi)/lambda).^(c-1)).
   *(exp((-del).*((th+2*k*pi)/lambda).^c)));
end
r=1+g;
x=r.*cos(th);
y=r.*sin(th);
plot(x,y,'g')
hold on
c=4.0;
lambda=2.0;
del=1.0;
g=0;
for k=0:n
```

```
    g=g+(((c.*del)./lambda).*(((th+2*k*pi)/lambda).^(c-1)).
    *(exp((-del).*((th+2*k*pi)/lambda).^c)));
end
r=1+g;
x=r.*cos(th);
y=r.*sin(th);
plot(x,y,'k')
hold on
xlabel('r*cos(\theta)')
ylabel('r*sin(\theta)')
title('PDF of Wrapped New Weibull Pareto
Distribution(Circular Representation)')
legend('C=2.0,\lambda=2.0,\delta=3.0','C=2.5,\
lambda=2.0,\delta=2.5','C=3.0,\lambda=2.0,\
delta=2.0','C=3.5,\lambda=2.0,\delta=1.5','C=4.0,\
lambda=2.0,\delta=1.0')
```

3. Program for the graph of cdf of wrapped exponentiated inverted Weibull distribution:

```
th=0.000001:0.001:2*pi;
c=2;
lambda=0.5;
n=3;
G=0;
for k=0:n
    G=G+(exp(-((th+2*k*pi).^(-c)).
    *lambda)-exp(-((2*k*pi).^(-c)).*lambda));
end
plot(th,G,'m')
hold on
c=2;
lambda=1.0;
G=0;
for k=0:n
    G=G+(exp(-((th+2*k*pi).^(-c)).
    *lambda)-exp(-((2*k*pi).^(-c)).*lambda));
end
plot(th,G,'c')
hold on
c=2;
lambda=1.5;
G=0;
for k=0:n
    G=G+(exp(-((th+2*k*pi).^(-c)).
    *lambda)-exp(-((2*k*pi).^(-c)).*lambda));
end
plot(th,G,'b')
hold on
```

```
c=2;
lambda=2.0;
G=0;
for k=0:n
    G=G+(exp(-((th+2*k*pi).^(-c)).
    *lambda)-exp(-((2*k*pi).^(-c)).*lambda));
end
plot(th,G,'k')
hold on
c=2;
lambda=2.5;
G=0;
for k=0:n
    G=G+(exp(-((th+2*k*pi).^(-c)).
    *lambda)-exp(-((2*k*pi).^(-c)).*lambda));
end
plot(th,G,'g')
hold off
xlabel('\theta')
ylabel('G(\theta)')
title('CDF Wrapped Exponentiated Inverted Weibull
Distribution')
legend('c=2.0,\lambda=0.5','c=2.0,\lambda=1.0','c=2.0,\
lambda=1.5','c=2.0,\lambda=2.0','c=2.0,\lambda=2.5')
```

4. Program for the graph of characteristic function of wrapped new Weibull–Pareto distribution:

```
x=[0.1377934705,0.7294545495,1.8083429027,3.4014336979,5.
5524961400,8.3301527468,11.8437858379,16.2792578314,21.9
965858120, 29.9206970123];
w=[0.3084411158,0.4011199292,0.2180682976,0.0620874561,0.009501
5170, 0.00075300839,0.000028259233,0.00000042493140,0.000000
0018395648, 0.0000000000009];
c=1.0;
lambda=1.0;
d=1.0;
t=linspace(0,100,101);
phi=ones(size(t));
i=find(t~=0);
phi(i)=w*exp(1i.*(lambda.*(x'./d).^(1/c))*t(i));
plot(t,real(phi),'-*b',t,imag(phi),'-*m')
ylabel('\phi(t)')
legend('real(\phi)','imaginary(\phi)')
xlabel('t')
title('Characteristic Function of Wrapped New Weibull
Pareto Distribution with c = 1.0, \lambda=1.0, \
delta=1.0')
```

5. Program for the population characteristics of wrapped new Weibull–Pareto distribution:

```
clc
x=[0.1377934705,0.7294545495,1.8083429027,3.4014336979,5.5
524961400,8.3301527468,11.8437858379,16.2792578314,21.9965858120,
29.9206970123];w=[0.3084411158,0.4011199292,0.2180682976,
0.0620874561,  0.0095015170,0.00075300839,0.000028259233,
0.00000042493140, 0.0000000018395648,0.0000000000009];
c=2.0;
lambda=2.0
d=3.0;
t=linspace(0,100,101);
phi=ones(size(t));
i=find(t~=0);
phi(i)= w*exp(1i.*(lambda.*(x'./d).^(1/c))*t(i));
    fprintf('\nPopulation Characteristics of New Weibull
    Pareto Distribution\n');
    fprintf('for C= %.2f',c);
    fprintf(' delta= %.2f',d);
    fprintf(' and lambda= %.2f are:\n' ,lambda);
    fprintf('\nTrigonometric Moments\n')
    alpha=[real(phi(2)),real(phi(3))]
    beta=[imag(phi(2)),imag(phi(3))]
    rho=sqrt((beta(1)).^2+(alpha(1)).^2);
   for i=1:2
   if beta(i)>=0 && alpha(i)>0
          mu(i)=atan(beta(i)/alpha(i));
   end
   if alpha(i)<0
          mu(i)=atan(beta(i)/alpha(i))+pi;
   end
   if beta(i)<0 && alpha(i)>=0
          mu(i)=atan(beta(i)/alpha(i))+2*pi;
   end
   if beta(i)>0 && alpha(i)==0
          mu(i)=pi/2;
   end
   end
    rho1=sqrt((beta).^2+(alpha).^2);
    p=1:2;
    alphag(p)=rho1.*cos(mu-p*mu(1));
    betag(p)=rho1.*sin(mu-p*mu(1));
    v0=1-rho(1);
    sig(1)=sqrt(abs(log(1/((rho1(1))^2))));
    sig(2)=sqrt(abs(log(1/((rho1(2))^2))));
    gamma1=(betag(2))./(v0.^(3/2));
    gamma2=(alphag(2)-((1-v0).^4))./(v0.^2);
    fprintf('Resultant Length ')
    rho1
```

```
    fprintf('Mean Direction %.4f \n',mu(1));
    fprintf('Circular Variance %.4f \n',v0)
    fprintf('Circular Standard Deviation :')
    p=1:2;
    disp(sig(p))
    fprintf('Central Trignometric Moments alpha1 & alpha2: \n')
    p=1:2;
    disp(alphag(p))
    fprintf('Central Trignometric Moments beta1 & beta2: \n')
    p=1:2;
    disp(betag(p))
    fprintf('Skewness is %.4f \n', gamma1)
    fprintf('Kurtosis is %.4f \n\n', gamma2)
```

6. Program for the graph of pdf of stereographic semicircular exponentiated inverted Weibull distribution:

```
clc
th=linspace((0+0.001),(pi-0.001),10000);
v=1.0;
c=2.0;
lambda=1.0;
g=(v*(sec(th/2)).^2.*lambda.*c.*(v.*tan
(th/2)).^(-(c+1)).*(exp(-(v.*tan(th/2)).^(-c)).^lambda))./2;
plot(th,g,'c')
hold on
c=2.0;
lambda=2.0;
g=(v*(sec(th/2)).^2.*lambda.*c.*(v.*tan
(th/2)).^(-(c+1)).*(exp(-(v.*tan(th/2)).^(-c)).^lambda))./2;
plot(th,g,'g')
hold on
c=2.0;
lambda=3.0;
g=(v*(sec(th/2)).^2.*lambda.*c.*(v.*tan
(th/2)).^(-(c+1)).*(exp(-(v.*tan(th/2)).^(-c)).^lambda))./2;
plot(th,g,'b')
hold on
c=2.0;
lambda=4.0;
g=(v*(sec(th/2)).^2.*lambda.*c.*(v.*tan
(th/2)).^(-(c+1)).*(exp(-(v.*tan(th/2)).^(-c)).^lambda))./2;
plot(th,g,'m')
xlabel('\theta')
ylabel('g(\theta)')
title('Graph of pdf of Stereographic EIW Distribution ')
legend('C=2.0,\lambda=1.0','C=2.0,\lambda=2.0','C=2.0,\
lambda=3.0','C=2.0,\lambda=4.0')
```

7. Program for the graph of cdf of stereographic semicircular new Weibull–Pareto distribution:

```
delta -
th=linspace((0+0.001),(pi-0.001),10000);
c=3.0;
lambda=3.0;
delta=3.0;
v=1;
G= 1-(exp(-delta.*((v.*tan(th/2))./lambda).^(c)));
G
plot(th,G,'c')
hold on
c=3.0;
lambda=4.0;
delta=2.0;
G= 1-(exp(-delta.*((v.*tan(th/2))./lambda).^(c)));
plot(th,G,'g')
hold on
c=4.0;
lambda=2.0;
delta=3.0;
G= 1-(exp(-delta.*((v.*tan(th/2))./lambda).^(c)));
plot(th,G,'b')
hold on
c=5.0;
lambda=3.0;
delta=4.0;
G= 1-(exp(-delta.*((v.*tan(th/2))./lambda).^(c)));
plot(th,G,'m')
xlabel('\theta')
ylabel('G(\theta)')
title('Graph of CDF of Stereographic Semicircular NWP
Distribution ')
legend('C=3.0,\lambda=3.0,\delta=3.0','C=3.0,\lambda=4.0,\
delta=2.0','C=4.0,\lambda=2.0,\delta=3.0','C=5.0,\lambda=3.0,\
delta=4.0')
```

8. Program for data plot for data set1 (movement of turtles):

```
clc;
t=linspace(0,2*pi,1000);
x=cos(t);
y=sin(t);
plot(x,y)
hold on
th=[8   9   13  13  14  18  22  27  30  34  38  38  40
44  45  47  48  48  48  48  50  53  56  57  58  58  61
63  64  64  64  65  65  68  70  73  78  78  78  83  83
```

```
88  88  88  90  92  92  93  95  96  98  100 103 106 113
118 138 153 153 155 204 215 223 226 237 238 243 244 250
251 257 268 285 319 343 350];
th
r=th.*pi./180;
x1=cos(r);
y1=sin(r);
plot(x1,y1,'*r')
hold on
th1=[13,38,58,65,83,92,153];
r2=th1.*pi./180;
 k2=[1,1.05];
 for i=1:2
x2=k2(i).*cos(r2);
y2=k2(i).*sin(r2);
plot(x2,y2,'*r')
 end
hold on
th3=[64,78,88];
r3=th3.*pi./180;
k3=[1,1.05,1.1];
for m3=1:3
x3=k3(m3).*cos(r3);
y3=k3(m3).*sin(r3);
plot(x3,y3,'*r')
end
th4=[48];
r4=th4.*pi./180;
k4=[1,1.05,1.1,1.15];
for m4=1:4
x4=k4(m4).*cos(r4);
y4=k4(m4).*sin(r4);
plot(x4,y4,'*r')
end
title('Orientations of turtles after laying eggs Data
Plot (n=76)')
axis equal;
```

9. Function for mle of wrapped exponentiated inverted Weibull distribution:

```
function [fun]=w_eiw_fminmle(X)
th=[8   9   13  13  14  18  22  27  30  34  38  38  40
44  45  47  48  48  48  48  50  53  56  57  58  58  61
63  64  64  64  65  65  68  70  73  78  78  78  83  83
88  88  88  90  92  92  93  95  96  98  100 103 106 113
118 138 153 153 155 204 215 223 226 237 238 243 244 250
251 257 268 285 319 343 350];
th=th.*pi./180;
```

```
th=sort(th);
n=76;
c=X(1)
lambda=X(2)
m=5;
g=0;
for k=0:m
    g=g+((lambda.*c).*((th+2*k*pi)).^(-(c+1))).
    *(exp(-(th+2*k*pi).^(-c)).^lambda);
end
fun=0;
for i=1:n
    fun=fun+log(g(i));
end
fun=-(fun);
return
```

10. Program invoking `fmincon` for mle of parameters of wrapped exponentiated inverted Weibull distribution:

```
clc
clear all;
X0=[1.0,1.0];
lb=[0;0];
n=76;
fun1=0;
[X,fval]=fmincon(@(X)W_eiw_fminmle(X),X0,[],[],[],[],lb,[])
maxlogL=-fval
```

11. Program for goodness of fit for stereographic semicircular new Weibull–Pareto model:

```
n=60;
th=[1   1   2   2   3   8   9   12  16  17  19  23  28  28  34
34  35  36  36  37  41  45  49  50  51  53  58  68  69  70  72
72  76  78  80  85  97  97  99  101 105 121 125 126 133 141
143 149 152 156 160 163 167 168 170 171 172 174 175 176];
th=th.*pi./180;
th=sort(th);
c=0.5797;
lambda=2.0848;
delta=1.0751;
G= 1-(exp(-delta.*((tan(th/2))./lambda).^(c)));
fprintf('\n1. Kuiper test\n');
R=sort(G);
d1=ones(size(n));
d2=ones(size(n));
for i=1:n
    d1(i)=(i/n)-R(i);
    d2(i)=R(i)-((i-1)/n);
```

```
end
dplus=max(d1);
dminus=max(d2);
vv=(dplus+dminus);
vk=sqrt(n)*vv;
v1=vk*(1+(0.155/sqrt(n))+(0.24/n))
fprintf('\n2. Watson s U2 test\n');
s2=sort(R);
i=1:n;
W2=sum((s2(i)-((i-0.5)/n)).^2)+(1/(12*n))
Wnew=4*(vv^2)/(pi^2);
```

12. Program for selection of a better model stereographic semicircular exponentiated inverted Weibull model:

```
clc;
%AIC, BIC of 76 Turtles data
n=76;
%AIC and BIC
MLL1=-119.5338
l1=MLL1;
AIC1=2*2-2*l1
BIC1=-2*l1+2*log(n)
%%%%%%%%%%%%%%%%%%%%%%%%%%%%%%%%%%%%%%%%%
fprintf('\nWrapped New Weibull Pareto Distribution\n')
MLL2=-118.7418
l2=MLL2;
AIC2=2*3-2*l2
BIC2=-2*l2+3*log(n)
```

13. Program for finding estimates of CR, ACR, and CCR angles for wrapped exponentiated inverted Weibull model at different sample sizes:

```
clc
clear all;
tic;
c=input('enter c = ');
lambda=input('enter lambda = ');
ssmax=100;
ss=5;
q=[5,10,10,10,10,10,10,10,10,10,10];
qq=0;
alpha=0.05;
iter=50;
k=1000;
fileID = fopen('RepQC_weiw_sampsize.txt','w');
fprintf(fileID,'Wrapped Exponentiated Inverted Weibull
Distribution\n');
```

```
fprintf(fileID,'with C = %.2f ',c);
fprintf(fileID,' lambda = %.2f\n',lambda);
fprintf(fileID,'Level of Significance =%.2f',alpha);
fprintf(fileID,'  Simulation size = %.0f',k);
fprintf(fileID,'  No of Iterations= %.0f \n\n',iter);
fprintf('\n for Wrapped Exponentiated Inverted Weibull
Distribution with\n C = %.2f ',c);
fprintf('lambda = %.2f\n',lambda);
fprintf('Level of Significance = %.2f',alpha);
fprintf(' Sample size = %.0f',ss);
fprintf(' and No of Iterations = %.0f\n',iter);
x=0.000001:0.001:2*pi;
g=0;
n1=5;
for j=0:n1
    g=g+((lambda.*c).*((x+2*j*pi)).^(-(c+1))).
    *(exp(-(x+2*j*pi). ^(-c)).^lambda);
end
%plot(x,g,'m');
v=max(g);
%fprintf('\nthe Maximum Value of Distribution at given
Parameters is %.4f:\n\n',v);
fprintf(fileID,'\n %6s %8s %8s %8s %8s %8s %8s\n','SampleS
ize','CMCR','CVCR','CMACR','CVACR','CMCCR','CVCCR');
while ss<=ssmax
    fprintf(' Sample size = %.0f\n',ss);
    %fprintf(fileID,'  Sample size = %.0f\n',ss);
    mmth=ones(size(iter));
    acr=ones(size(iter));
    ccr=ones(size(iter));
    itn=1;
    n= ss;
    %fprintf(fileID,'%8s %6s  %9s  %9s \n','Iteration','cr',
    'acr','ccr');
    while itn<=iter
        %fprintf('\n Iteration No = %.0f\n',itn);
        ctr=0;
        mth=ones(size(k));
        while ctr<k
            ktr=0;
            th=ones(size(n));
            u=ones(size(2));
            while ktr<n
                for i=1:2
                    u(i)=rand;
                end
                u(1);
                u(2);
                z=2*pi*u(1);
                fz=0;
```

```
                for j=0:n1
                   fz=fz+((lambda.*c).*((z+2*j*pi)).^(-(c+1))).
                   *(exp(-(z+2*j*pi).^(-c)).^lambda);
                end
                y=v*u(2);
                fz;
                if (y<=fz)
                    y;
                    fz;
                    z;
                    r=z;
                    ktr=ktr+1;
                    th(ktr)=r;
                end
            end
            th;
            s=0;
            t=0;
            for i=1:n
                s=s+sin(th(i));
                t=t+cos(th(i));
            end
            s=s/n;
            t=t/n;
            ctr=ctr+1;
            if s>=0 && t>0
                mth(ctr)=atan(s/t);
            end
            if t<0
                mth(ctr)=atan(s/t)+pi;
            end
            if s<0 && t>=0
                mth(ctr)=atan(s/t)+ 2*pi;
            end
            if s>0 && t==0
                mth(ctr)=pi/2;
            end
            %fprintf('\nFor the sample %.0f ',ctr)
            %fprintf('\n  CR in radians is %.4f',mth(ctr));
            %fprintf(' and in degrees is %.2f\n\n ',
            (mth(ctr)*180)/pi);
        end
        s=0;
        t=0;
        for i=1:k
            s=s+sin(mth(i));
            t=t+cos(mth(i));
        end
        s=s/k;
        t=t/k;
```

```
        if s>=0 && t>0
            mmth(itn)=atan(s/t);
        end
        if t<0
            mmth(itn)=atan(s/t)+pi;
        end
        if s<0 && t>=0
            mmth(itn) =atan(s/t)+ 2*pi;
        end
        if s>0 && t==0
            mmth(itn) =pi/2;
        end
        orderth=sort(mth);
        q1=k-round((alpha/2)*k);
        h1=round((alpha/2)*k)+1;
        acr(itn) =orderth(q1);
        ccr(itn) =orderth(h1);
        %fprintf('\n  CR in radians is %.4f',mmth(itn));
        %fprintf(' and in degrees is %.2f\n ',
        (mmth(itn)*180)/pi);
        %fprintf('ACR in radians is %.4f',acr(itn));
        %fprintf(' and in degrees is %.2f\n ',
        (acr(itn)*180)/   pi);
        %fprintf('CCR in radians is %.4f',ccr(itn));
        %fprintf(' and in degrees is %.2f\n ',
        (ccr(itn)*180)/pi);
        %op=[itn,mmth(itn),acr(itn),ccr(itn)];
        %fprintf(fileID,'%5.0f      %.4f      %.4f      %.4f
        \r\n',op);
        itn=itn+1;
    end
    sacr=0;
    tacr=0;
    sccr=0;
    tccr=0;
    for i=1:iter
        sacr=sacr+sin(acr(i));
        tacr=tacr+cos(acr(i));
        sccr=sccr+sin(ccr(i));
        tccr=tccr+cos(ccr(i));
    end
    sacr=sacr/iter;
    tacr=tacr/iter;
    sccr=sccr/iter;
    tccr=tccr/iter;
    if sacr>=0 && tacr>0
        cmacr =atan(sacr/tacr);
    end
    if tacr<0
        cmacr =atan(sacr/tacr)+pi;
```

```
        end
        if sacr<0 && tacr>=0
            cmacr =atan(sacr/tacr)+ 2*pi;
        end
        if sacr>0 && tacr==0
            cmacr =pi/2;
        end
        cmacrd=cmacr*180/pi();
        if sccr>=0 && tccr>0
            cmccr =atan(sccr/tccr);
        end
        if tccr<0
            cmccr =atan(sccr/tccr)+pi;
        end
        if sccr<0 && tccr>=0
            cmccr =atan(sccr/tccr)+ 2*pi;
        end
        if sccr>0 && tccr==0
            cmccr =pi/2;
        end
        cmccrd=cmccr*180/pi;
        s=0;
        t=0;
        for i=1:iter
            s=s+sin(mmth(i));
            t=t+cos(mmth(i));
        end
        s=s/iter;
        t=t/iter;
        if s>=0 && t>0
            cmmth =atan(s/t);
        end
        if t<0
            cmmth =atan(s/t)+pi;
        end
        if s<0 && t>=0
            cmmth =atan(s/t)+ 2*pi;
        end
        if s>0 && t==0
            cmmth =pi/2;
        end
        cmmthd=cmmth*180/pi;
        cvacr=0;
        cvccr=0;
        cvth=0;
        for i=1:iter
            cvacr=cvacr+(1-cos(acr(i)-cmacr));
            cvccr=cvccr+(1-cos(ccr(i)-cmccr));
            cvth=cvth+(1-cos(mmth(i)-cmmth));
        end
```

```
        cvacr=cvacr/iter;
        cvacrd=cvacr*180/pi;
        cvccr=cvccr/iter;
        cvccrd=cvccr*180/pi;
        cvth=cvth/iter;
        cvthd=cvth*180/pi;
        op1=[ss,cmmthd,cvthd,cmacrd,cvacrd,cmccrd,cvccrd];
        fprintf(fileID,'%6.0f       %9.2f %7.2f  %7.2f %7.2f
        %7.2f  %7.2f \r\n',op1);
        fprintf('\n  CMCR in radians is %.4f',cmmth);
        fprintf(' and in degrees is %.2f\n ',cmmthd);
        fprintf(' CVCR in radians is %.4f',cvth);
        fprintf(' and in degrees is %.2f\n ',cvthd);
        fprintf('CMACR in radians is %.4f',cmacr);
        fprintf(' and in degrees is %.2f\n ',cmacrd);
        fprintf('CVACR in radians is %.4f',cvacr);
        fprintf(' and in degrees is %.2f\n ',cvacrd);
        fprintf('CMCCR in radians is %.4f',cmccr);
        fprintf(' and in degrees is %.2f\n ',cmccrd);
        fprintf('CVCCR in radians is %.4f',cvccr);
        fprintf(' and in degrees is %.2f\n\n ',cvccrd);
        qq=qq+1;
        ss=ss+q(qq);
    end
    toc;
```

14. Program for finding estimates of CR, ACR, and CCR angles for wrapped new Weibull–Pareto model at different simulation sizes:

```
    clc
    tic;
    c=input('enter c = ');
    del= input('enter delta = ');
    lambda=input('enter lambda = ');
    ss=5;
    alpha=0.05;
    iter=50;
    simax=10000;
    q=[250,500,1000,1000,1000,1000,1000,1000,1000,1000,1000,
    1000, 1000];
    qq=0;
    k=250;
    fileID = fopen('RepQC_wnwp_Simulation.txt','w');
    fprintf(fileID,'Simulation for Wrapped New Weibull Pareto
    Distribution\n');
    fprintf(fileID,'with C = %.2f ',c);
    fprintf(fileID,'delta = %.2f',del);
    fprintf(fileID,' lambda = %.2f\n',lambda);
    fprintf(fileID,'Level of Significance =%.2f',alpha);
    fprintf(fileID,'  Sample size = %.0f ',ss);
```

```
fprintf(fileID,'  No of Iterations= %.0f \n\n',iter);
fprintf('\nSimulation for Wrapped New Weibull Pareto
Distribution with\n C = %.2f ',c);
fprintf('delta = %.2f  ',del);
fprintf('lambda = %.2f\n',lambda);
fprintf('Level of Significance = %.2f',alpha);
fprintf(' Sample size = %.0f',ss);
fprintf(' and No of Iterations = %.0f\n',iter);
x=0.000001:0.001:2*pi;
n1=5;
g=0;
for j=0:n1
    g=g+(((c.*del)./lambda).*(((x+2*j*pi)/lambda).^(c-1)).
    *(exp((-del).*((x+2*j*pi)/lambda).^c)));
end
%plot(x,g,'m');
v=max(g);
%fprintf('\nthe Maximum Value of Distribution at given
Parameters is %.4f:\n\n',v);
fprintf(fileID,'\n %10s %9s %9s %8s %8s %8s
%8s\n','SimuSize',' CMCR',' CVCR',' CMACR',' CVACR','
CMCCR','CVCCR');
while k<=simax
    fprintf(' Simulation size = %.0f\n',k);
    mmth=ones(size(iter));
    acr=ones(size(iter));
    ccr=ones(size(iter));
    itn=1;
    n= ss;
    %fprintf(fileID,'%8s %6s  %9s  %9s
    \n','Iteration','cr','acr','ccr');
    while itn<=iter
        %fprintf('\n Iteration No = %.0f\n',itn);
        ctr=0;
        mth=ones(size(k));
        while ctr<k
            ktr=0;
            th=ones(size(n));
            u=ones(size(2));
            while ktr<n
                for i=1:2
                    u(i)=rand;
                end
                u(1);
                u(2);
                z=2*pi*u(1);
                fz=0;
                for j=0:n1
```

```
fz=fz+(((c.*del)./lambda).*(((z+2*j*pi)/lambda).^(c-
1)).*(exp((-del).*((z+2*j*pi)/lambda).^c)));
                end
                y=v*u(2);
                fz;
                if (y<=fz)
                    y;
                    fz;
                    z;
                    r=z;
                    ktr=ktr+1;
                    th(ktr)=r;
                end
            end
            th;
            s=0;
            t=0;
            for i=1:n
                s=s+sin(th(i));
                t=t+cos(th(i));
            end
            s=s/n;
            t=t/n;
            ctr=ctr+1;
            if s>=0 && t>0
                mth(ctr)=atan(s/t);
            end
            if t<0
                mth(ctr)=atan(s/t)+pi;
            end
            if s<0 && t>=0
                mth(ctr)=atan(s/t)+ 2*pi;
            end
            if s>0 && t==0
                mth(ctr)=pi/2;
            end
            %fprintf('\nFor the sample %.0f ',ctr)
            %fprintf('\n  CR in radians is %.4f',mth(ctr));
            %fprintf(' and in degrees is %.2f\n\n ',
            (mth(ctr) *180)/pi);
        end
        s=0;
        t=0;
        for i=1:k
            s=s+sin(mth(i));
            t=t+cos(mth(i));
        end
        s=s/k;
        t=t/k;
        if s>=0 && t>0
```

```
            mmth(itn)=atan(s/t);
        end
        if t<0
            mmth(itn)=atan(s/t)+pi;
        end
        if s<0 && t>=0
            mmth(itn) =atan(s/t)+ 2*pi;
        end
        if s>0 && t==0
            mmth(itn) =pi/2;
        end
        orderth=sort(mth);
        q1=k-round((alpha/2)*k);
        h1=round((alpha/2)*k)+1;
        acr(itn) =orderth(q1);
        ccr(itn) =orderth(h1);
        %fprintf('\n  CR in radians is %.4f',mmth(itn));
        %fprintf(' and in degrees is %.2f\n ',
        (mmth(itn)*180)/pi);
        %fprintf('ACR in radians is %.4f',acr(itn));
        %fprintf(' and in degrees is %.2f\n ',
        (acr(itn)*180)/pi);
        %fprintf('CCR in radians is %.4f',ccr(itn));
        %fprintf(' and in degrees is %.2f\n ',
        (ccr(itn)*180)/pi);
        %op=[itn,mmth(itn),acr(itn),ccr(itn)];
        %fprintf(fileID,'%5.0f  %.4f  %.4f
        %.4f \r\n',op);
        itn=itn+1;
    end
    sacr=0;
    tacr=0;
    sccr=0;
    tccr=0;
    for i=1:iter
        sacr=sacr+sin(acr(i));
        tacr=tacr+cos(acr(i));
        sccr=sccr+sin(ccr(i));
        tccr=tccr+cos(ccr(i));
    end
    sacr=sacr/iter;
    tacr=tacr/iter;
    sccr=sccr/iter;
    tccr=tccr/iter;
    if sacr>=0 && tacr>0
        cmacr =atan(sacr/tacr);
    end
    if tacr<0
        cmacr =atan(sacr/tacr)+pi;
    end
```

```
if sacr<0 && tacr>=0
    cmacr =atan(sacr/tacr)+ 2*pi;
end
if sacr>0 && tacr==0
    cmacr =pi/2;
end
cmacrd=cmacr*180/pi();
if sccr>=0 && tccr>0
    cmccr =atan(sccr/tccr);
end
if tccr<0
    cmccr =atan(sccr/tccr)+pi;
end
if sccr<0 && tccr>=0
    cmccr =atan(sccr/tccr)+ 2*pi;
end
if sccr>0 && tccr==0
    cmccr =pi/2;
end
cmccrd=cmccr*180/pi;
s=0;
t=0;
for i=1:iter
    s=s+sin(mmth(i));
    t=t+cos(mmth(i));
end
s=s/iter;
t=t/iter;
if s>=0 && t>0
    cmmth =atan(s/t);
end
if t<0
    cmmth =atan(s/t)+pi;
end
if s<0 && t>=0
    cmmth =atan(s/t)+ 2*pi;
end
if s>0 && t==0
    cmmth =pi/2;
end
cmmthd=cmmth*180/pi;
cvacr=0;
cvccr=0;
cvth=0;
for i=1:iter
    cvacr=cvacr+(1-cos(acr(i)-cmacr));
    cvccr=cvccr+(1-cos(ccr(i)-cmccr));
    cvth=cvth+(1-cos(mmth(i)-cmmth));
end
cvacr=cvacr/iter;
```

```
        cvacrd=cvacr*180/pi;
        cvccr=cvccr/iter;
        cvccrd=cvccr*180/pi;
        cvth=cvth/iter;
        cvthd=cvth*180/pi;
        op1=[k,cmmthd,cvthd,cmacrd,cvacrd,cmccrd,cvccrd];
        fprintf(fileID,'   %6.0f %.2f %.2f  %.2f  %.2f  %.2f
        %.2f \r\n',op1);
        fprintf('\n  CMCR in radians is %.4f',cmmth);
        fprintf(' and in degrees is %.2f\n ',cmmthd);
        fprintf(' CVCR in radians is %.4f',cvth);
        fprintf(' and in degrees is %.2f\n ',cvthd);
        fprintf('CMACR in radians is %.4f',cmacr);
        fprintf(' and in degrees is %.2f\n ',cmacrd);
        fprintf('CVACR in radians is %.4f',cvacr);
        fprintf(' and in degrees is %.2f\n ',cvacrd);
        fprintf('CMCCR in radians is %.4f',cmccr);
        fprintf(' and in degrees is %.2f\n ',cmccrd);
        fprintf('CVCCR in radians is %.4f',cvccr);
        fprintf(' and in degrees is %.2f\n\n ',cvccrd);
        qq=qq+1;
        k=k+q(qq);
    end
    toc;
```

15. Program for plotting CR, ACR, and CCR angles:

```
clc
clear all
cr= input( 'CR in degrees=');
acr= input( 'ACR in degrees=');
ccr= input( 'CCR in degrees=');
x0=0;
y0=0;
cr=cr*pi/180;
x1=cos(cr);
y1=sin(cr);
plot([x0 x0+x1],[y0,y0+y1],'b');
hold on
acr=acr*pi/180;
x2=cos(acr);
y2=sin(acr);
plot([x0 x0+x2],[y0,y0+y2],'r');
hold on
ccr=ccr*pi/180;
x3=cos(ccr);
y3=sin(ccr);
plot([x0 x0+x3],[y0,y0+y3],'g');
hold on
angle= linspace(0,2*pi);
```

```
x=cos(angle);
y=sin(angle);
plot(x,y,'m');
axis equal
hold off
xlabel('Cos(\theta)')
ylabel('Sin(\theta)')
title('Plotting of CR, ACR and CCR  (in degrees)')
cr1=num2str(cr*180/pi,'%.2f');
acr1=num2str(acr*180/pi,'%.2f');
ccr1=num2str(ccr*180/pi,'%.2f');
legend(['CR=',cr1,'°'],['ACR=',acr1,'°'],['CCR=',
ccr1,'°'])
```

A2.2 Population Characteristics (Tables A2.1 through A2.4)

TABLE A2.1

Population Characteristics of WEIWD at $\lambda = 2$

c	$c = 0.5$	$c = 1.5$	$c = 2.0$	$c = 2.5$	$c = 3.0$
Trigonometric Moments					
α_1	−0.0279	0.3009	−0.1147	−0.1098	−0.0063
α_2	−0.1353	−0.1451	−0.4282	−0.2263	−0.4357
β_1	0.5497	0.5067	0.4467	0.7085	0.8347
β_2	0.5425	−0.2179	0.4915	0.1864	0.0940
Resultant Length					
ρ_1	0.5504	0.5893	0.4612	0.7170	0.8347
ρ_2	0.5591	0.2618	0.6519	0.2932	0.4457
Mean Direction					
μ_0	1.6215	1.0349	1.8222	1.7246	1.5783
Variance					
v_0	0.4496	0.4107	0.5388	0.2830	0.1653
Standard Deviation					
σ_0	1.0928	1.0284	1.2441	0.8157	0.6011
	1.0783	1.6371	0.9250	1.5665	1.2712
Central Trigonometric Moments					
α_1^*	0.5504	0.5893	0.4612	0.7170	0.8347
α_2^*	0.0797	−0.1220	0.1383	0.1593	0.4343
β_1^*	0.0000	0.0000	0.0000	0.0000	0.0000
β_1^*	−0.5534	0.2317	−0.6371	−0.2461	−0.1005

(*Continued*)

TABLE A2.1 (*Continued*)

Population Characteristics of WEIWD at $\lambda = 2$

c	$c = 0.5$	$c = 1.5$	$c = 2.0$	$c = 2.5$	$c = 3.0$
Skewness					
γ_1^0	−1.8358	0.8803	−1.6109	−1.6349	−1.4960
Kurtosis					
γ_2^0	−0.0596	−1.4384	0.3207	−1.3110	−1.8765

TABLE A2.2

Population Characteristics of WEIWD at $c = 2$

λ	$\lambda = 0.5$	$\lambda = 1.5$	$\lambda = 2.0$	$\lambda = 2.5$	$\lambda = 3.0$
Trigonometric Moments					
α_1	0.4262	−0.0583	−0.1147	−0.1133	−0.0862
α_2	−0.1147	−0.1268	−0.4282	−0.6898	−0.7536
β_1	0.7225	0.5645	0.4467	0.3626	0.3181
β_2	0.4467	0.4844	0.4915	0.2556	−0.1030
Resultant Length					
ρ_1	0.8388	0.5675	0.4612	0.3799	0.3296
ρ_2	0.4612	0.5007	0.6519	0.7357	0.7606
Mean Direction					
μ_0	1.0378	1.6738	1.8222	1.8737	1.8353
Variance					
ν_0	0.1612	0.4325	0.5388	0.6201	0.6704
Standard Deviation					
σ_0	0.5928	1.0645	1.2441	1.3912	1.4899
	1.2441	1.1762	0.9250	0.7835	0.7398
Central Trigonometric Moments					
α_1^*	0.8388	0.5675	0.4612	0.3799	0.3296
α_2^*	0.4465	0.0251	0.1383	0.4215	0.7026
β_1^*	0.0000	0.0000	0.0000	0.0000	0.0000
β_1^*	−0.1156	−0.5001	−0.6371	−0.6030	−0.2914
Skewness					
γ_1^0	−1.7875	−1.7582	−1.6109	−1.2349	−0.5308
Kurtosis					
γ_2^0	−1.8729	−0.4203	0.3207	1.0420	1.5370

TABLE A2.3

Population Characteristics of SSNWPD at $\lambda = 3$ and $\delta = 3$

c	$c = 2.0$	$c = 3.0$	$c = 4.0$	$c = 5.0$	$c = 6.0$
Trigonometric Moments					
α_1	−0.2318	−0.4640	−0.5740	−0.6328	−0.6683
α_2	−0.4753	−0.4181	−0.2757	−0.1649	−0.0862
β_1	0.8510	0.8311	0.7894	0.7561	0.7316
β_2	−0.3288	−0.7000	−0.8630	−0.9303	−0.9603
Resultant Length					
ρ_1	0.8820	0.9518	0.9760	0.9860	0.9910
ρ_2	0.5780	0.8154	0.9060	0.9448	0.9641
Mean Direction					
μ_0	1.8367	2.0799	2.1995	2.2676	2.3110
Variance					
v_0	0.1180	0.0482	0.0240	0.0140	0.0090
Standard Deviation					
σ_0	0.5010	0.3142	0.2205	0.1678	0.1348
	1.0471	0.6389	0.4445	0.3371	0.2703
Central Trigonometric Moments					
α_1^*	0.8820	0.9518	0.9760	0.9860	0.9910
α_2^*	0.5764	0.8153	0.9059	0.9448	0.9641
β_1^*	0.0000	0.0000	0.0000	0.0000	0.0000
β_1^*	0.0423	0.0114	0.0038	0.0016	0.0008
Skewness					
γ_1^0	1.0444	1.0828	1.0184	0.9532	0.8989
Kurtosis					
γ_2^0	−2.0747	−2.3482	−2.4262	−2.4460	−2.4476

TABLE A2.4

Population Characteristics of SSNWPD at $c = 3$ and $\delta = 3$

λ	$\lambda = 2.0$	$\lambda = 3.0$	$\lambda = 4.0$	$\lambda = 5.0$	$\lambda = 6.0$
Trigonometric Moments					
α_1	−0.1395	−0.4640	−0.6443	−0.7506	−0.8170
α_2	−0.7376	−0.4181	−0.0837	0.1751	0.3631
β_1	0.9302	0.8311	0.7173	0.6197	0.5408
β_2	−0.2377	−0.7000	−0.8569	−0.8785	−0.8459
Resultant Length					
ρ_1	0.9406	0.9518	0.9641	0.9733	0.9798
ρ_2	0.7749	0.8154	0.8610	0.8958	0.9205
Mean Direction					
μ_0	1.7196	2.0799	2.3027	2.4514	2.5568
Variance					
v_0	0.0594	0.0482	0.0359	0.0267	0.0202
Standard Deviation					
σ_0	0.3500	0.3142	0.2702	0.2325	0.2021
	0.7141	0.6389	0.5471	0.4691	0.4070
Central Trigonometric Moments					
α_1^*	0.9406	0.9518	0.9641	0.9733	0.9798
α_2^*	0.7749	0.8153	0.8610	0.8958	0.9205
β_1^*	0.0000	0.0000	0.0000	0.0000	0.0000
β_1^*	0.0110	0.0114	0.0084	0.0057	0.0039
Skewness					
γ_1^0	0.7564	1.0828	1.2329	1.3110	1.3561
Kurtosis					
γ_2^0	−2.2216	−2.3482	−2.4269	−2.4786	−2.5131

Index

O

P

Q

R

S